ABOUT THIS WORKBOOK

Basic Technical Drawing Student Workbook is intended primarily for use with *Basic Technical Drawing* (Eighth Edition, ©2004) by Spencer, Dygdon, and Novak. All references and instructions refer to that text. However, this workbook may also be used with other reference texts.

Since the time available in many schools for technical drawing has become increasingly limited, the primary objective has been to provide in as few sheets as possible a complete and thorough coverage of the fundamentals. It is expected that in many cases the instructor will supplement these problem sheets with assignments of problems from the text to be drawn on blank paper, vellum, or film.

Many of the problems are based upon actual industrial designs. Their presentations conform with the latest American National Drafting Standards. Less space is devoted to lettering and other routine exercises intended primarily to develop skills. Considerably more emphasis is given to technical sketching. Numerous problems in the book make use of grids similar to the various cross-section papers available commercially. A special effort has been made to present problems that are thought-provoking, rather than merely requiring a great deal of routine drafting.

Since many of the problems in this workbook and the accompanying textbook are of a general nature, they can also be solved on most computer-aided drafting (CAD) systems. If a CAD system is available, the instructor may choose to assign specific problems to be completed by this method.

In line with the increased interest in charts and graphs, intersections and developments, and descriptive geometry, a number of problems in these areas have been provided.

All sheets are 8 $\frac{1}{2}$" x 11", in conformity with the American National Drafting Standards. This size also facilitates handling and filing by the instructor. In general a 4H pencil (0.3 mm fine-line mechanical) will be found suitable for construction lines and guide lines for lettering; a 2H pencil (0.3 mm) for center lines, section lines, dimension lines, extension lines, and phantom lines; and an F pencil (0.7 mm) for general line-work and lettering. The student should make all construction lines very light, and *he or she should not erase them.*

The instructions, which start on page 5, provide detailed information together with references to the text for each problem. The student is urged to study these instructions and references carefully before starting each problem.

The authors wish to express appreciation to their colleagues for many valuable suggestions and to the numerous industrial firms who have so generously cooperated in supplying problem material. Comments and criticism for users of this problems book will be most welcome.

JOHN THOMAS DYGDON

JAMES E. NOVAK

Contents

Basic Technical Drawing Student Workbook
Copyright © by Glencoe/McGraw-Hill.

Instructions

References are to the Eighth Edition of *Basic Technical Drawing* (©2004) by Spencer, Dygdon, and Novak.

DRAWING A-1. VERTICAL CAPITALS AND NUMERALS

References: Secs. 5.1-5.11. All lettering on this sheet and the remaining sheets in this workbook is to be done *freehand*. The necessary guide lines, both horizontal and vertical, for Drawings A-1, A-2, and A-3 have already been drawn. Note that the horizontal guide lines control the heights of the characters and that the randomly spaced vertical guide lines serve simply as an aid in keeping the characters uniformly vertical.

- **Exercise 1.** Using a medium sharp HB pencil, letter the characters shown in the spaces provided. These letters and numerals may first be sketched lightly and then corrected where necessary, before being made clean-cut and dark using the strokes shown. Omit the arrows and numbers from your letters.
- **Exercise 2.** Using a fairly sharp F pencil, reproduce the given words and numerals in the spaces directly below each line. Pay particular attention to the spacing of letters and words.
- **Exercise 3.** Proceed as in Exercise 2. In the title strip under DRAWN BY, letter your name, with the last name first, using a sharp F pencil. Under FILE NO., letter the identification symbol assigned by your instructor.

DRAWING A-2. VERTICAL CAPITALS AND NUMERALS

References: Secs. 5.4-5.11, 10.2-10.11. All necessary guide lines are provided. All lettering must be made clean-cut and dark.

- **Exercises 1-5.** Using a sharp F pencil, duplicate on the right side the information shown, including arrowheads and finish marks.
- **Exercise 6.** Reproduce the title shown in the space provided at the right. Pay particular attention to the accurate centering of each line of lettering in the title, about the given vertical center line.

DRAWING A-3. VERTICAL LOWERCASE LETTERING

References: Secs. 5.7, 5.10, 5.14. All necessary guide lines have been provided. Note the additional horizontal guide lines for lowercase lettering. All lettering must be made clean-cut and dark.

- **Exercise 1.** Using a medium sharp HB pencil, letter the characters shown in the spaces provided. Use the strokes shown, but omit the arrows and numbers from your letters.
- **Exercise 2.** Using a fairly sharp F pencil, reproduce the given words in the spaces directly below each line. Pay particular attention to the spacing of letters and words.
- **Exercise 3.** Proceed as in Exercise 2; use a sharp F pencil.

DRAWING B-1. INCLINED CAPITALS AND NUMERALS

References: Secs. 5.1-5.7, 5.9, 5.10, 5.12, 5.13. All lettering on this sheet and the remaining sheets in this workbook is to be done *freehand*. The necessary guide lines, both horizontal and inclined, for Drawings B-1, B-2, and B-3 have already been drawn. Note that the horizontal guide lines control the height of the characters and that the randomly-spaced inclined guide lines serve simply as an aid in keeping the characters uniformly inclined.

- **Exercise 1.** Using a medium sharp HB pencil, letter the characters shown in the spaces provided. These letters and numerals may first be sketched lightly, and then corrected where necessary, before being made clean-cut and dark using the strokes shown. Omit the arrows and numbers from your letters.
- **Exercise 2.** Using a fairly sharp F pencil, reproduce the given words and numerals in the spaces directly below each line. Pay particular attention to the spacing of letters and words.
- **Exercise 3.** Proceed as in Exercise 2. In the title strip under DRAWN BY, letter your name with the last name first, using a sharp F pencil.

Under FILE NO., letter the identification symbol assigned by your instructor.

DRAWING B-2. INCLINED CAPITALS AND NUMERALS

References: Secs. 5.4-5.7, 5.9, 5.10, 5.12, 5.13, 10.2-10.11. All necessary guide lines are provided. All lettering must be made clean-cut and dark.

- **Exercises 1-5.** Using a sharp F pencil, duplicate on the right side the information shown, including arrowheads and finish marks.
- **Exercise 6.** Reproduce the title shown in the space provided at the right. Pay particular attention to the accurate centering of each line of lettering in the title, about the given vertical center line.

DRAWING B-3. INCLINED LOWERCASE LETTERING

References: Secs. 5.7, 5.10, 5.15. All necessary guide lines have been provided. Note the additional horizontal guide lines for lowercase lettering. All lettering must be clean-cut and dark.

- **Exercise 1.** Using a medium sharp HB pencil, letter the characters shown in the spaces provided. Use the strokes shown, but omit the arrows and numbers from your letters.
- **Exercise 2.** Using a fairly sharp F pencil, reproduce the given words in the spaces directly below each line. Pay particular attention to the spacing of letters and words.
- **Exercise 3.** Proceed as in Exercise 2; use a sharp F pencil.

DRAWING C-1. USE OF T-SQUARE AND TRIANGLES

References: Secs. 4.1-4.14, 4.17, 4.18. Before fastening this sheet and subsequent sheets to the drawing surface, align the sheet horizontally by adjusting the sheet so that one of the principal horizontal lines coincides with the upper edge of the T-square. Fasten the sheet to the drawing surface with drafting tape.

- **Spaces 1, 3, 5, and 7.** Draw *thick* parallel lines beginning at the starting marks. Use a sharp F pencil, and go over the lines until the thickness matches that of the printed lines. The lines in Space 1 are drawn horizontal. The inclination of lines in Space 3 is 45° with horizontal, upward to the right. In Space 5 the inclination is 30° with horizontal, upward to the right. In Space 7 the inclination is 60° with horizontal, upward to the right.
- **Spaces 2, 4, 6, and 8.** Draw *thin* parallel lines beginning at the starting marks. Use a sharp 2H pencil to draw these lines, which should be just as dark as the thick lines, but very thin. The lines in Space 2 are drawn vertical. The inclination of lines in Space 4 is 45° with horizontal, downward to the right. In Space 6 the inclination is 30° with horizontal, downward to the right. In Space 8 the inclination is 60° with horizontal, downward to the right.

  In the title strip, using a 4H pencil, draw very light horizontal guide lines from the starting marks under DRAWN BY and random vertical or inclined guide lines as assigned. Using a sharp F pencil, letter your name with the last name first. Under FILE NO., letter the identification symbol assigned by your instructor.

DRAWING C-2. ALPHABET OF LINES

References: Secs. 4.10, 4.14-4.16.

- **Space 1.** Beginning at the starting marks, reproduce each of the given lines in the space directly below each line. Use an F pencil for the thick and medium weight lines and a 2H pencil for the thin lines.
- **Space 2.** Beginning at the starting marks, draw parallel visible lines at 15° with horizontal, upward to the right.
- **Space 3.** Beginning at the starting marks, draw parallel hidden lines at 75° with horizontal, downward to the right.

- **Space 4.** Beginning at the starting marks and using the T-square and triangle, draw section lines parallel to the given lines.
- **Space 5.** Beginning at the starting marks and using the T-square and triangle, draw extension lines perpendicular to the given center line.

DRAWING C-3. SCALES AND LAYOUT

References: Secs. 4.19-4.23, 4.33-4.35, 8.1-8.6.

- **Space 1.** Use the architects or engineers scale as necessary. Measure lines A through F at the scales indicated, and record the measured lengths in the appropriate spaces in the table provided. At G through M, draw lines of specified lengths at the scales shown. Terminate each line with a short, thin, vertical dash similar to the given lines above. At N through P, determine the scales and lengths of lines and record this information in the appropriate spaces in the table provided.
- **Spaces 2 and 3.** Using the starting corners and dimensions given, draw the views shown to the scales specified. Your final lines should be visible line weight. Omit all dimensions, extension lines, and dimension lines. Note that the scale in Space 2 is full-size and that the scale in Space 3 is half-size. Do not scale the reduced drawings.

DRAWING C-4. LAYOUT AND USE OF COMPASS

References: Secs. 4.24-4.29, 4.34, 4.35. Using the starting corners and dimensions given, draw the views shown to the scales specified. Your final lines should be visible line weight. Omit all dimensions, extension lines, dimension lines, and notes. Note that the scale in Spaces 1 and 2 is full-size and that the scale in Space 3 is half-size. Do not scale the reduced drawings.

DRAWING D-1. GEOMETRIC CONSTRUCTIONS

References: Secs. 6.1-6.3. All construction lines used to solve problems on this sheet are to be drawn very light (4H pencil) and are not to be erased. Add center lines where necessary.

- **Space 1.** References: Figs. 6-2, 6-3. Divide the V-rack so it will contain six equally spaced teeth. Start the first tooth at A as indicated.
- **Space 2.** References: Figs. 6-6, 6-14. Locate the draw hole as indicated.
- **Space 3.** Reference: Fig. 6-7. Locate and draw hole as indicated.
- **Space 4.** Reference: Fig. 6-9. Complete the view of the triangle in the new location.
- **Space 5.** References: Figs. 6-16, 6-17. Draw the notch as specified, and complete the view.
- **Space 6.** Reference Fig. 6-13 (c). Complete the view of the Square Head Wrench.

DRAWING D-2. GEOMETRIC CONSTRUCTIONS

References: Secs. 6.1-6.3. All construction lines used to solve problems on this sheet are to be drawn very light (4H pencil) and are not to be erased. Add center lines where necessary.

- **Space 1.** References: Fig. 6-13 (m) and (n). Complete the view of the traffic sign to the scale specified. Omit lettering.
- **Space 2.** References: Fig. 6.13 (f) and (g). Draw the view of the Hexagon Patio Block to the scale specified.
- **Space 3.** Reference: Fig. 6-7. Complete the view of the Cover Plate.
- **Space 4.** Reference: Fig. 6-23. Using the approximate ellipse method, complete the view of the Cam that has a major axis of $3\frac{1}{8}$" in length along the horizontal center line, and a minor axis of 2" in length along the vertical center line.
- **Space 5.** Reference: Fig. 6-22. Using the concentric-circle method, complete the view of the semi-elliptical Arch. Locate a minimum of 10 points to establish the curve. Using the irregular curve, draw a smooth dark curve through the plotted points.

Drawing D-3. Geometric Constructions

References: Secs. 6.1-6.3. All construction lines used to solve problems on this sheet are to be drawn very light (4H pencil) and are not to be erased. Add center lines where necessary and indicate all tangent points with light dashes.

- **Space 1.** Reference: Fig. 6-18. Complete the view of the Bracket.
- **Space 2.** References: Figs. 6-17, 6-18. Complete the view of the Index Finger.
- **Space 3.** References: Fig. 6-19 (a) and (b). Complete the view of the Hinge.
- **Space 4.** References: Fig. 6-19 (a) and (b). Complete the view of the Guide Arm.
- **Space 5.** Reference: Fig. 6-19 (c). Complete the view of the Rollers.
- **Space 6.** Reference: Fig. 6-19 (c). Complete the view of the Lever.

Drawing E-1. Two-View Drawings

References: Secs. 4.1, 4.10, 4.19, 4.24-4.29, 7.8, 8.1-8.7. Using the starting corners shown, reproduce the two-view drawings *full-size*. Final line weights should conform to the standard alphabet of lines. Omit dimensions. Do not scale the reduced drawings.

Drawing E-2. Three-View Drawings

References: Secs. 4.1, 4.10, 4.19, 4.24-4.29, 7.10, 8.1-8.6, 8.10. Using the starting corners shown, reproduce the three views of the Bracket to the scale indicated. Final line weights should conform to the standard alphabet of lines. Omit dimensions unless assigned. If dimensions are required, use $1/8$" high lettering, vertical or inclined, as assigned. Do not scale the drawing.

Drawing F-1. One- and Two-View Technical Sketching

References: Secs. 2.1-2.10, 7.9. All lines on this sheet are to be drawn freehand. Make all final lines clean-cut and dark.

- **Spaces 1 and 2.** Sketch indicated lines over given lines.

- **Spaces 3 and 4.** Starting at the given corners, sketch the views. Count squares on the small sketches in order to obtain the size of the views.
- **Space 5.** Starting at the given corner, sketch the two views. Make the spacing and size the same number of squares as in the small sketch.

Drawing F-2. Isometric Technical Sketching

References: Secs. 17.1-17.5. All lines on this sheet are to be drawn freehand. Make all final lines clean-cut and dark. Using the starting corners given, reproduce the two isometric sketches on the enlarged grid, counting squares to obtain the size.

Drawing F-3. Oblique Technical Sketching

References: Secs. 17.18, 17.19. All lines on this sheet are to be drawn freehand. Make all final lines clean-cut and dark. Using the starting corners given, reproduce the two oblique sketches on the enlarged grid, counting squares to obtain the size.

Drawing G-1. Identification of Surfaces

References: Secs. 7.1-7.5, 7.17, 8.11-8.13. In each of the three problems the various surfaces are identified by letters on the isometric drawing and by numbers on the three views. In the table, letter with a sharp F pencil the numbers corresponding to the letters given at the left, using the guide lines provided.

Drawing H-1. Six Principal Views

References: Secs. 7.3-7.7, 7.14-7.17, 8.11-8.15. Sketch the views as indicated. Make all lines clean-cut and dark. In the space provided, letter the names of the necessary views.

Drawing H-2. Six Principal Views

References: Secs. 7.3-7.7, 7.14-7.17, 8.11-8.15. Sketch the views as indicated. Make all lines clean-cut and dark. In the space provided, letter the names of the necessary views.

DRAWING H-3. SKETCHING VIEWS AND ISOMETRICS

References: Secs. 7.11, 7.17, 8.11-8.15, 17.3-17.5, 17.7. Sketch the views as indicated, spacing the principal views three squares apart. Make all final lines clean-cut and dark.

DRAWING H-4. SKETCHING VIEWS AND ISOMETRICS

References: Secs. 7.11, 7.17, 8.11-8.15, 17.3-17.5, 17.7. In each of the problems, a front and a right-side view and an incomplete isometric of an object are given. Study the two views and complete the isometric sketch. Then, using the starting corners given, add the top views. Make all final lines clean-cut and dark.

DRAWING H-5. MISSING LINES

References: Secs. 7.3-7.5, 7.11, 7.14-7.18, 8.8, 8.10-8.14. In each of the problems, lines (visible, hidden, or center lines) are missing from one or more views. Add all missing lines, freehand or with instruments, as assigned.

DRAWING H-6. MISSING LINES.

References: Secs. 7.3-7.5, 7.11, 7.14-7.18, 8.8, 8.10-8.15, 8.17. In each of the problems, lines (visible, hidden, or center lines) are missing from one or more views. Add all missing lines, freehand or with instruments as assigned.

DRAWING H-7. MISSING VIEWS

References: Secs. 7.3-7.5, 7.10, 7.11, 7.14-7.18, 8.8, 8.11-8.15, 8.17, 8.18. In each problem two complete views are given and a third view is missing. Add the third view in each case freehand.

DRAWING H-8. MISSING VIEWS

References: Secs. 7.3-7.5, 7.10, 7.11, 7.14-7.18, 8.8, 8.11-8.15, 8.17, 8.18, 10.7, 10.10, 11.3. In each problem two complete views are given and a third view is missing. Add the third view in each case, using instruments. In problem 7 the fillets and rounds are $1/_8$R and may be drawn freehand. Also, show finish marks in the front view.

DRAWING H-9. MISSING VIEWS

References: Secs. 7.3-7.5, 7.10, 7.11, 7.14-7.18, 8.8, 8.11-8.15, 8.17, 8.18. In each problem two complete views are given and a third view is missing. Add the third view in each case, using instruments.

DRAWING H-10. MISSING VIEWS

References: Secs. 7.3-7.5, 7.10, 7.11, 7.14-7.18, 8.8, 8.11-8.15, 8.17, 8.18. In each problem two complete views are given and a third view is missing. Add the third view in each case, using instruments.

DRAWING H-11. MISSING VIEWS

References: Secs. 7.3-7.5, 7.10, 7.11, 7.14-7.18, 8.8, 8.11-8.15, 8.17, 8.18. In each problem two complete views are given and a third view is missing. Add the third view in each case, using instruments.

DRAWING H-12. MISSING VIEWS

References: Secs. 7.3-7.5, 7.10, 7.11, 7.14-7.18, 8.8, 8.11-8.15, 8.17, 8.18, 10.7, 10.10, 11.3. In each problem two complete views are given and a third view is missing. Add the third view in each case, using instruments. In problem 3 the fillets and rounds are $1/_8$R and may be drawn freehand. Also, show finish marks in the top view.

DRAWING H-13. THREE-VIEW MECHANICAL DRAWING

References: Secs. 7.1-7.7, 7.10, 7.12, 7.14-7.18, 8.8-8.15, 8.17, 8.18, 10.7, 10.10, 11.3. Using instruments and the starting corners given, draw the front, top, and right side views of the Slide Bracket. Note that the scale of the drawing is *half-size*. Omit dimensions unless assigned. Show finish marks.

DRAWING J-1. DIMENSIONING

References: Secs. 10.1-10.25, 11.1-11.19. Add Dimensions, using instruments and spacing dimension lines $^3/_8$" from the views and $^3/_8$" apart. Use $^1/_8$" lettering and guide lines for figures and notes.

- **Problem 1.** Dimension the views, but use the letters *S* and *L*, instead of numerals, to represent size and location dimensions. The small hole is a drilled hole. Since the material is cold rolled steel (CRS), the object is understood to be finished all over (FAO). Finish marks are not necessary.
- **Problem 2.** Dimension the views completely, including finish marks. The two small holes are drilled. The scale is full-size.

DRAWING J-2. DIMENSIONING

References: Secs. 10.1-10.29, 11.1-11.19. Add dimensions, using instruments and spacing dimension lines $^3/_8$" from the views and $^3/_8$" apart. Use $^1/_8$" lettering and guide lines for figures and notes. Since the material specified for both problems is cold rolled steel (CRS), the objects are understood to be finished all over (FAO). Finish marks are not necessary.

- **Problem 1.** Add complete dimensions. For the hole, use the note: .749-.750 REAM. The scale is full-size.
- **Problem 2.** Add complete dimensions. The scale is half-size.

DRAWING J-3. DIMENSIONING

References: Secs. 10.1-10.29, 11.1-11.19. Add dimensions, using instruments and spacing dimension lines $^3/_8$" from the views and $^3/_8$" apart. Use $^1/_8$" lettering and guide lines for figures and notes. The fillets and rounds in both problems are $^1/_8$R.

- **Problem 1.** Dimension the views completely, including finish marks. For the circular hole, use the note: .812-.813 REAM.

- **Problem 2.** Dimension the views completely, including finish marks. The large central hole is drilled. Give the depth of this hole in the note. The two small holes are also drilled. The scale is half-size.

DRAWING J-4. DIMENSIONING

References: Secs. 10.1-10.29, 11.1-11.19. Add dimensions, using instruments and dimension lines $^3/_8$" from the views and $^3/_8$" apart. Use $^1/_8$" lettering and guide lines for figures and notes. The smaller hole in the center is reamed. Use the note: .749-.750 REAM. The large hole in the center is counterbored. Give the depth of the counterbore in the note. The small hole on the left side of the object is drilled. Fillets and rounds are $^1/_8$R (R.12). Add finish marks. The scale is full-size.

DRAWING K-1. SECTIONAL VIEWS

References: Secs. 12.1-12.4. Using an HB pencil, sketch full or half sections as indicated. Make section lines thin to contrast well with the heavy visible lines, spacing them approximately $^3/_{32}$" apart. Make all final lines clean-cut and dark so that the sketches will stand out from the grid lines. Omit cutting planes unless assigned.

DRAWING K-2. SECTIONAL VIEWS

References: Secs. 12.1-12.4. In each problem all sectioned-lined areas are shown completely. Add all missing lines, freehand or mechanically as assigned, to the sectioned views, including center lines. Omit hidden lines in the sectioned views.

DRAWING K-3. SECTIONAL VIEWS

References: Secs. 12.1-12.4. Draw the indicated sectional views, using instruments. In Problem 2 the fillets and rounds are $^1/_8$R. Add finish marks to all views in Problem 2. Omit hidden lines in the sectioned views. Use a sharp 2H pencil for all section lines, spacing them approximately $^3/_{32}$" apart.

DRAWING K-4. SECTIONAL VIEWS

References: Secs. 12.1-12.7, 12.12.

- **Space 1.** Draw revolved sections as indicated, using break lines on each side of the hexagonal section.
- **Space 2.** Draw revolved sections, with "S" breaks on each side of sections.
- **Space 3.** Draw removed sections including all visible lines behind the cutting plane in each case, and all necessary center lines. Omit hidden lines.

DRAWING K-5. SECTIONAL VIEWS

References: 12.1-12.6, 12.8-12.12. Draw the indicated sectional views, using instruments. In all four problems fillets and rounds are $1/8$R, and should be drawn freehand. Add finish marks in all problems. Problems 1 and 2 are full sections. Use a sharp 2H pencil for all section lines, spacing them approximately $3/32$" apart.

DRAWING L-1. PRIMARY AUXILIARY VIEWS

References: Secs. 13.1-13.6, 13.8. Sketch auxiliary views as indicated. Using an HB pencil, make visible and hidden lines dark so that the views will stand out clearly from the grids. Show and label reference planes in all problems. In Problems 1 and 2 the auxiliary views of only the inclined surfaces are required. In Problems 3 and 4 the auxiliary views are to be complete, including all hidden lines. In Problems 1 and 2, number the corners of surfaces A and B in the auxiliary views the same size as those given.

DRAWING L-2. PRIMARY AUXILIARY VIEWS

References: Secs. 13.1-13.6, 13.8-13.11. Add all missing lines, freehand or mechanically as assigned, in the regular or auxiliary views.

DRAWING L-3. PRIMARY AUXILIARY VIEWS

References: Secs. 13.1-13.6. Using instruments, draw complete auxiliary views as indicated. Show all hidden lines and center lines. Show and label reference planes in all problems.

DRAWING L-4. PRIMARY AUXILIARY VIEWS

References: Secs. 13.1-13.7, 13.9. Using instruments, draw complete auxiliary views as indicated. Show all hidden lines and center lines. Show and label reference planes in all problems. In Problem 2, dimension the angle between surfaces A and B (measure with protractor after the auxiliary view has been completed). In Problem 4, plot at least seven points for each curve in the auxiliary view. Using the irregular curve, draw a smooth curve through the plotted points.

DRAWING L-5. PRIMARY AND SECONDARY AUXILIARY VIEWS

References: Secs. 13.1-13.6, 13.8, 13.9, 13.11, 13.12. Using instruments, draw the indicated auxiliary views. Show all hidden lines and center lines. Show and label reference planes in each problem. In Problem 2, dimension the 140° angle in the primary auxiliary view.

DRAWING M-1. PRIMARY REVOLUTIONS

References: Secs. 14.1-14.5, 14.8. Using instruments, draw the indicated views. Show all hidden lines. In each problem, letter the answers to the questions, using the guide lines provided.

- **Problem 1.** At the left are given the front and top views of an object to be revolved. At the right the front view has been revolved counterclockwise. Draw the resulting top and right-side views.
- **Problem 2.** At the left are given the front and top views of an object to be revolved. At the right the top view has been revolved counterclockwise. Draw the resulting front and right side views.
- **Problem 3.** At the left are given the front, top, and right-side views of an object to be revolved. At the right, the right-side view has been revolved counterclockwise. Draw the resulting front and top views.

- **Problem 4.** At the left are given the front and top views of an object to be revolved. In the center the top view has been revolved clockwise until the line of intersection between surfaces A and B will appear as a point in the right-side view. Draw the resulting front and right-side views, and dimension the angle between surfaces A and B.

DRAWING M-2. PRIMARY AND SUCCESSIVE AUXILIARY VIEWS

References: Secs. 14.1-14.8. Using instruments, draw the indicated views. Show all hidden lines.

- **Problem 1.** At the left are given two views in which one revolution has been already accomplished. In the center the object is further revolved about a different axis. Draw the resulting front and right-side views. Remember the rule: *If lines are parallel on the object, they will be parallel in any view of the object.*
- **Problem 2.** The top view of an object which has been revolved clockwise is shown. At the upper right is shown a reduced scale drawing of the object in the unrevolved position, with certain necessary dimensions given. Draw the resulting front and right-side views, showing all construction.

DRAWING N-1. THREAD NOMENCLATURE AND IDENTIFICATION

References: Secs. 15.1-15.18. Using the guide lines given, letter the appropriate information or thread notes in the spaces provided. Where space is limited, use standard abbreviations. Each of the drawings is made to scale, full-size, unless otherwise indicated. Information not obvious from each drawing is given immediately below that drawing.

DRAWING N-2. DETAILED UNIFIED THREADS

References: Secs. 15.1-15.9, 15.12, 15.17, 15.18.

- **Problem 1.** Draw detailed external, internal, and mating Unified threads as indicated. Add necessary section lining and complete the view. Complete the leader and add arrowhead touching the threads. The scale of the drawing is full-size.
- **Problem 2.** Draw detailed Unified threads as indicated. Complete the leaders and add arrowheads touching the threads. The scale of the drawing is full-size.

DRAWING N-3. DETAILED SQUARE AND ACME THREADS

References: Secs. 15.10, 15.11, 15.18. Draw detailed external, internal, and mating Square and Acme threads as indicated. Add necessary section lining and complete the views. Complete the leaders and add arrowheads touching the threads. The scale of the drawing in Problem 1 is half-size, and in Problem 2 is full-size.

DRAWING N-4. SCHEMATIC THREADS AND FASTENERS

References: Secs. 15.13-15.18, 15.20-15.22; Appendix Table 7.

- **Problem 1.** Complete the sectioned assembly by drawing the specified thread details, using the schematic thread symbols. Add necessary section lining and complete the views. Complete the leaders and add arrowheads where required. Chamfer ends of external threads 45° x thread depth.
- **Problem 2.** Draw the two bolts and nuts as indicated, using the schematic thread symbol. Add necessary section lining and complete the views.

DRAWING N-5. STANDARD FASTENERS

References: Secs. 15.23-15.30, Appendix Tables 7, 8, 9, 10, 11, and 14.

- **Problem 1.** Complete the sectioned assembly as indicated by drawing the specified fasteners. Add necessary section lining and complete the view.

- **Problem 2.** Using the guide lines provided, letter the complete name of each of the items shown. Standard abbreviations may be used to avoid crowding. Omit size specifications.

DRAWING O-1. ISOMETRIC DRAWING

References: Secs. 17.1, 17.2, 17.6-17.13. Using instruments and the starting corners indicated, make full-size isometric drawings. Show all construction. In Problems 1 and 2, use dividers to transfer all measurements from the views to the isometrics. In Problem 3, the views are not to scale and you must use the given dimensions to make the isometric drawing. Omit hidden lines.

DRAWING O-2. ISOMETRIC DRAWING

References: Secs. 17.1, 17.2, 17.6-17.15. Using instruments and the starting corners indicated, make full-size isometric drawings. Show all construction. Note that in each of the problems the views are not to scale, and you must use the given dimensions to make the isometric drawings. Omit hidden lines.

DRAWING O-3. ISOMETRIC DRAWING

References: Secs. 17.1, 17.2, 17.6-17.16. Using instruments and the starting corners indicated, make full-size isometric drawings. Show all construction. Omit hidden lines.

- **Problem 1.** Use dividers to transfer all measurements from the views to the isometric. Plot points for the large curve, as shown in Fig. 17-16, and use the irregular curve.
- **Problem 2.** The views are not to scale. Use the given dimensions to make the isometric drawing. Alternate assignments: Make isometric drawing in half-section or full-section as assigned.

DRAWING O-4. ISOMETRIC DRAWING

References: Secs. 17.1, 17.2, 17.6-17.15. Using instruments and the starting corners indicated, make full-size isometric drawings. Show all

construction. Note that in each of the problems the views are not to scale, and you must use the given dimensions to make the isometric drawings. Omit hidden lines.

DRAWING P-1. OBLIQUE DRAWING

References: Secs. 17.20-17.25. Using instruments and the starting corners indicated, make full-size cavalier drawings. Show all construction including points of tangency. Note that in each of the problems the views are not to scale, and you must use the given dimensions to make the oblique drawings. Omit hidden lines and center lines.

DRAWING P-2. OBLIQUE DRAWING

References: Secs. 17.20-17.26. Using instruments and the starting corners indicated, make full-size cavalier or cabinet drawings as specified. Show all construction including points of tangency. Note that in each of the problems the views are not to scale, and you must use the given dimensions to make the oblique drawings. Problem 1 is a cavalier drawing in half-section. Omit hidden lines and center lines.

DRAWING P-3. OBLIQUE DRAWING

References: Secs. 17.20-17.25. Using instruments and the starting corners indicated, make full-size cabinet or cavalier drawings as specified. Show all construction including points of tangency. Note that in each problem the views are not to scale, and you must use the given dimensions to make the oblique drawings. In Problem 2, plot points for the elliptical curve, and use the irregular curve. Omit hidden lines and center lines.

DRAWING Q-1. PARALLEL LINE DEVELOPMENTS

References: Secs. 18.1-18.9. Draw developments *inside up*, so that all bending lines will be on the inside of the object and not seen after it is folded into shape or *fabricated*. It is customary to start or finish developments with the shortest edge or element to obtain a minimum seam length.

- **Problem 1.** The front and top views of a truncated hexagonal prism are given. Construct the development, beginning at edge A-1 shown to the right of the front view. Do not include the top or bottom surfaces (bases) of the prism. Identify end points of the edges on the development.
- **Problem 2.** Given are the front and top views of a right circular cylinder that is cut at the top and bottom. Construct the development of the vertical surface of the cylinder only, starting at element A-1. Divide the circle in the top view into 16 divisions. Omit bases.

DRAWING Q-2. RADIAL LINE DEVELOPMENTS

References: Secs. 18.1-18.3, 18.13-18.19. Draw developments *inside up*, so that all bending lines will be on the inside of the object and not seen after it is folded into shape or *fabricated*. It is customary to start or finish developments with the shortest edge or element to obtain a minimum seam length.

- **Problem 1.** The front and incomplete top views of a truncated pyramid are given. Complete the top view. Extend given lines where necessary to join the lines of intersection. Construct the development beginning with edge A-1, which is to be set off along a horizontal line to the left of point V. Develop the complete pyramid (not truncated) first, and then construct the intersection for the cut. Omit bases. Identify end points of the edges on the development.
- **Problem 2.** The front and top views of a truncated oblique cone are given. Construct the *half* development beginning with edge A-1, which is to be set off along a horizontal line to the left of point V. Develop half the cone first, and construct the curve for the cut next. Omit bases. Divide the upper or lower half of the base in the top view into 8 divisions. Identify all elements in the given views and the development.

DRAWING Q-3. INTERSECTIONS OF PRISMS AND PYRAMID

References: Secs. 18.22, 18.23, 18.25, 18.27.
- **Problem 1.** The top, right side, and incomplete front views of intersecting prisms are given. Complete the front view, drawing the lines of intersection. Extend given lines, where necessary, to join the lines of intersection. Show visibility. Identify points on the lines of intersection. If assigned, developments of each prism should be drawn on a separate sheet of paper.
- **Problem 2.** The top and incomplete front and right side views of an intersecting pyramid and prism are given. Complete the front and right side views, drawing the lines of intersection. Extend given lines, where necessary, to join the lines of intersection. Show visibility. Identify points on the lines of intersection. If assigned, developments of the prism and pyramid should be drawn on a separate sheet of paper.

DRAWING Q-4. INTERSECTIONS OF CYLINDERS AND CONE

References: Secs. 18.22, 18.24-18.26, 18.28, 18.29.
- **Problem 1.** The right-side and incomplete front and top views of two intersecting cylinders are given. Complete the front and top views, drawing the curve of intersection. Use the irregular curve. Show visibility. Show all construction. Extend given lines, where necessary, to complete the views. Identify at least three points on the curve of intersection. If assigned, developments of each cylinder should be drawn on a separate sheet of paper.
- **Problem 2.** The right side and incomplete front and top views of an intersecting cone and cylinder are given. Complete the front and top views, drawing the curve of intersection. Use the irregular curve. Show visibility. Show all construction. Extend given lines, where necessary, to complete the views. Identify at least three points on the curve of intersection.

If assigned, developments of the cylinder and cone should be drawn on a separate sheet of paper.

DRAWING R-1. PIE AND BAR CHARTS

References: Secs. 19.1-19.3, 19.5.

- **Problem 1.** 1997 passenger-car production in the United States was distributed as follows:

Chrysler Corp.	441,000
Ford Motor Co.	1,290,000
General Motors Corp.	2,270,000
Honda Motor Co.	648,000
Toyota Motor Co.	554,000
Others	565,000

Divide the pie chart to illustrate the above data. Show and label the appropriate percentages and automobile production for each manufacturer. Use $1/8$" engineering lettering throughout. Balance the largest sector symmetrically about a vertical center line in the upper area of the circle. Title the chart: U.S. PASSENGER CAR PRODUCTION—1997.

- **Problem 2.** Construct a bar chart showing the population (in millions) of the State of Illinois for the years indicated.

Year	Population
1930	7.6
1940	7.9
1950	8.7
1960	10.1
1970	11.1
1980	11.4
1990	11.4

Use $1/4$" wide horizontal bars beginning at the left vertical axis. Space the bars $1/4$" apart. Cross-hatch each bar with 45° section lines approximately $3/32$" apart. Draw the vertical grid lines (thin) from the main scale divisions, but not across the bars. Title the chart: POPULATION OF ILLINOIS. Subtitle: 1930-1990. Use 1/8" engineering lettering for the title.

DRAWING R-2. LINE CHARTS

References: Secs. 19.1, 19.4.

- **Problem 1.** The normal monthly temperatures for Chicago, Illinois, are as follows:

Jan. 26°	Feb. 28°	Mar. 36°	Apr. 49°
May 60°	June 71°	July 76°	Aug. 74°
Sept. 66°	Oct. 55°	Nov. 40°	Dec. 29°

Make a line chart showing this information. Draw horizontal and vertical grid lines (thin) for the major scale divisions. Title the chart: CHICAGO NORMAL MONTHLY TEMPERATURES. Use $1/8$" engineering lettering for the title.

- **Problem 2.** The most home runs hit by an individual baseball player in the major leagues for the period 1987 to 1998 for the American League and the National League were as follows:

YEAR	HOME RUNS AMERICAN LEAGUE	HOME RUNS NATIONAL LEAGUE
1987	49	49
1988	42	39
1989	36	47
1990	51	40
1991	44	38
1992	43	35
1993	46	46
1994	40	43
1995	50	40
1996	52	47
1997	56	49
1998	56	70

Make a line chart comparing the most home runs hit by individual players in the American and National leagues. Draw thin horizontal and vertical grid lines for the major scale divisions. Use a solid line to represent the American League. Use a dashed line to represent the National League. Identify each line. Title the chart: MAJOR LEAGUE HOME RUN LEADERS. Subtitle the chart: 1987-1998. Use $1/8$" engineering lettering for the title.

DRAWING S-1. POINTS, LINES, AND PLANES

References: Secs. 7.3, 7.5, 7.10, 8.11-8.14, 13.2-13.6, 13.9, 13.11, 25.1. Show and label reference planes in all problems. Use the reference planes as bases for transferring distances. All lettering is to be 0.10 high.

- **Problem 1.** The solution to this problem is given. Study the example carefully and note the use, placement, and representation of the reference planes. Note that all points in the right side view are the same perpendicular distance from reference plane X-X as they are in the top view.

- **Problem 2.** The top and front views of line 1-2 are given. Draw the right side view of line 1-2, using reference planes in the top and right side views.

- **Problem 3.** The top and front views of line 1-2 are given. Draw an auxiliary view of line 1-2 looking in the direction of arrow A, using reference planes in the top and auxiliary views.

- **Problem 4.** The top and front views of points 1, 2, and 3 are given. Actually these three points represent a plane 1-2-3. Draw the right side view of points 1, 2, and 3, using reference planes in the top and right side views.

- **Problem 5.** The top and front views of plane 1-2-3 are given. Draw the right side view of plane 1-2-3, using reference planes in the top and right side views.

- **Problem 6.** The top and front views of intersecting lines 1-2 and 3-4 are given. Actually these intersecting lines represent a plane 1-2-3-4. Draw the right side view of intersecting lines 1-2 and 3-4, using reference planes in the top and right side views.

DRAWING S-2. TRUE LENGTH OF LINE AND TRUE SIZE OF PLANES

References: Secs. 13.6, 13.11, 25.2, 25.3. Show and label reference planes in all problems. All lettering is to be 0.10 high.

- **Problem 1.** The top and front views of line 1-2 are given. Draw the auxiliary view looking in the direction of the arrow, which is perpendicular to line 1-2 in the top view. The resulting auxiliary view will show line 1-2 in *true length* (TL), and the inclination of the line with the horizontal plane or *grade*. Show and label the true length and grade on the drawing *and* in the spaces provided.

- **Problem 2.** The top and front views of line 1-2 are given. Draw the auxiliary view looking in the direction of the arrow, which is perpendicular to line 1-2 in the front view. The resulting auxiliary view will show line 1-2 in true length (TL), and the inclination of line 1-2 with the frontal plane. Show and label the true length and angle with the frontal plane on the drawing *and* in the spaces provided.

- **Problem 3.** The top and front views of plane 1-2-3 are given. First, draw an auxiliary view showing the *edge view* (EV) of plane 1-2-3. Then draw a secondary auxiliary view showing the *true size* (TS) of the plane. Show and label the edge view and true size of the plane on the drawing.

DRAWING S-3. PIERCING POINT OF LINE AND PLANE, AND INTERSECTION OF PLANE AND PRISM

References: Secs. 13.6, 13.11, 25.2-25.4. Show and label reference planes in each problem. All lettering is to be 0.10 high.

- **Problem 1.** Find the piercing point of line 4-5 and plane 1-2-3. Indicate the piercing point by a small circle in each view. Show proper visibility in the given views.

- **Problem 2.** Draw the intersection of unlimited plane 4-5-6 and the prism. Show proper visibility in all views.

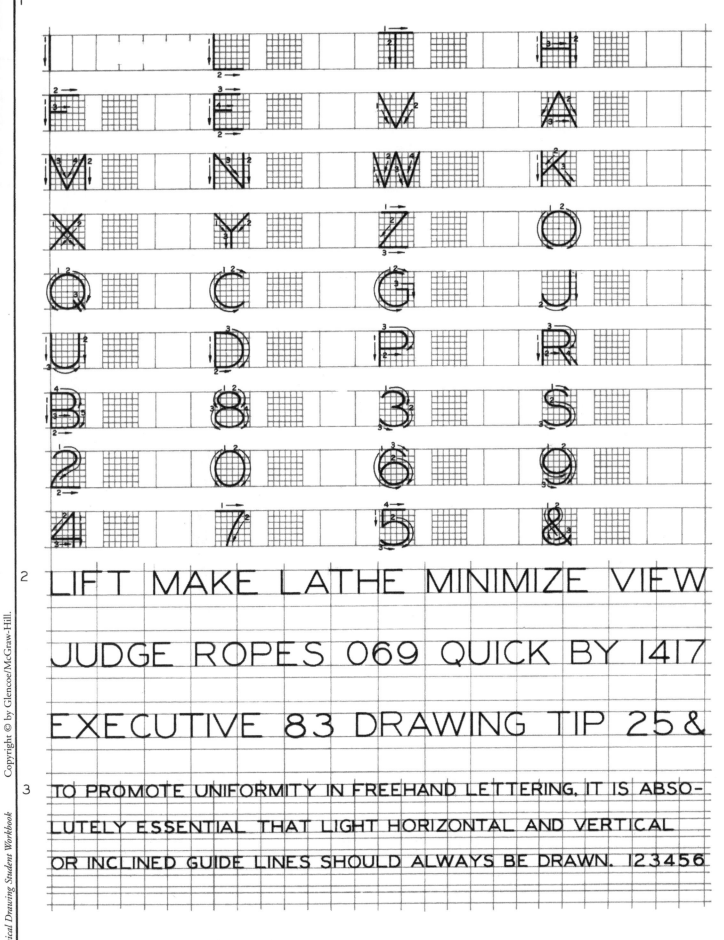

LIFT MAKE LATHE MINIMIZE VIEW

JUDGE ROPES 069 QUICK BY 1417

EXECUTIVE 83 DRAWING TIP 25 &

TO PROMOTE UNIFORMITY IN FREEHAND LETTERING, IT IS ABSO-
LUTELY ESSENTIAL THAT LIGHT HORIZONTAL AND VERTICAL
OR INCLINED GUIDE LINES SHOULD ALWAYS BE DRAWN. 123456

CAPITALS AND NUMERALS	DRAWN BY	FILE NO.	DRAWING
VERTICAL LETTERING			**A-1**

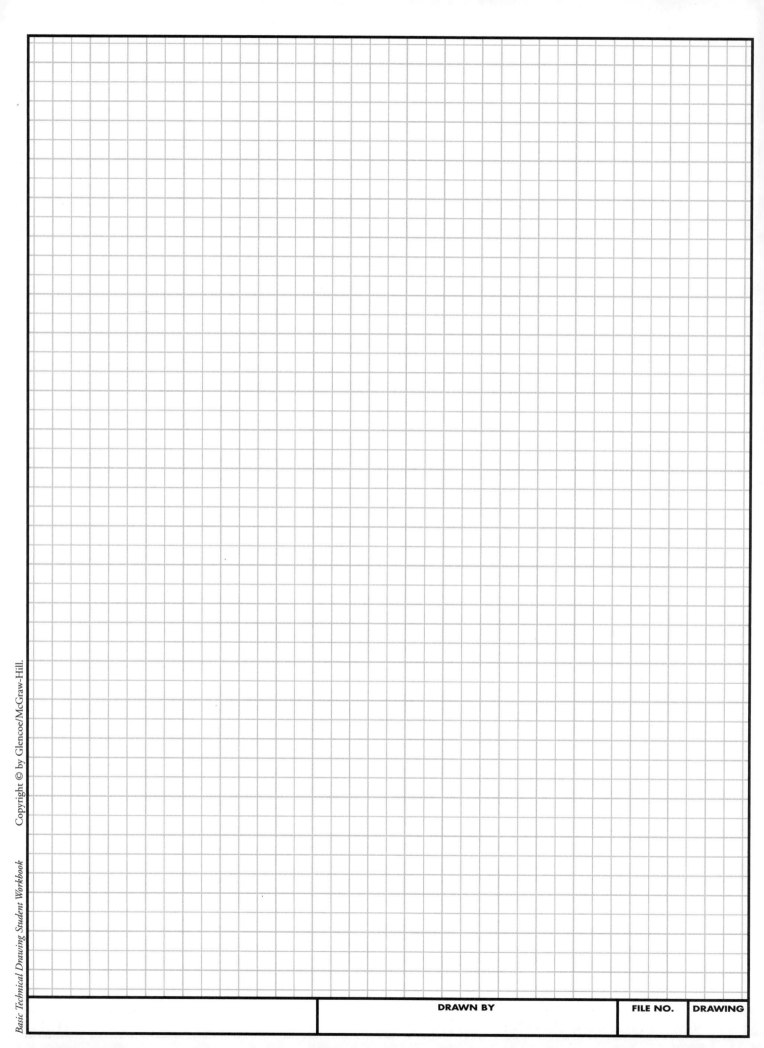

		DRAWN BY	FILE NO.	DRAWING

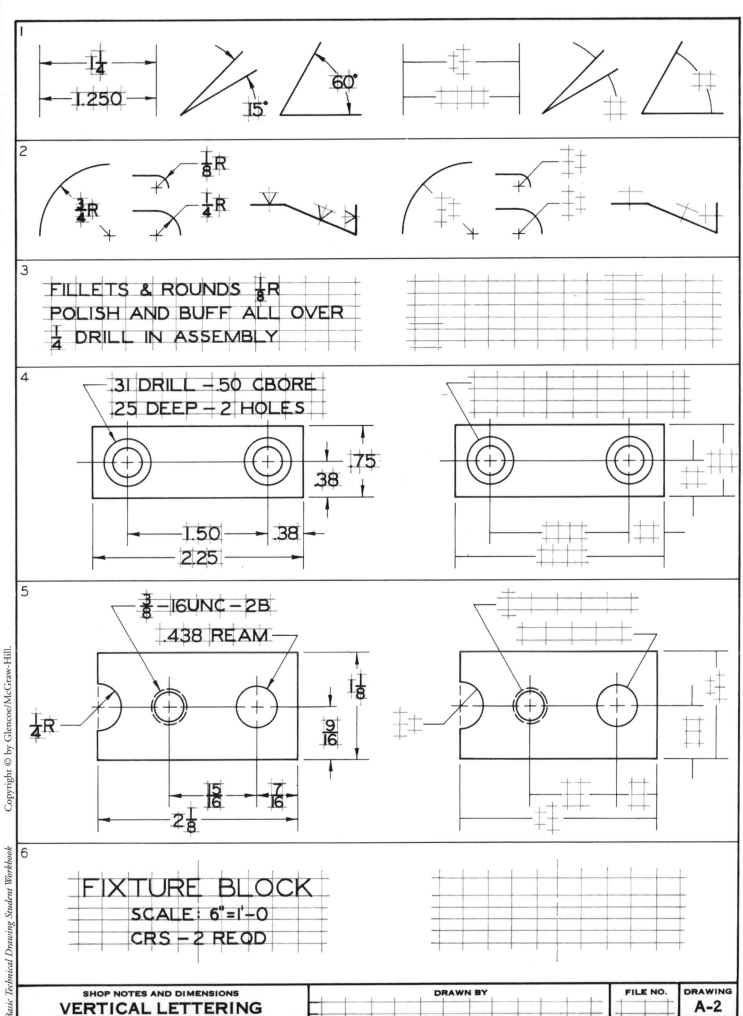

1

$1\frac{1}{4}$

1.250

15°

60°

2

$\frac{1}{8}$R

$\frac{3}{4}$R

$\frac{1}{4}$R

3

FILLETS & ROUNDS $\frac{1}{8}$R
POLISH AND BUFF ALL OVER
$\frac{1}{4}$ DRILL IN ASSEMBLY

4

31 DRILL –.50 CBORE
25 DEEP – 2 HOLES

.75

.38

1.50

.38

2.25

5

$\frac{3}{8}$ –16UNC –2B

.438 REAM

$\frac{1}{4}$R

$1\frac{1}{8}$

$\frac{9}{16}$

$\frac{15}{16}$

$\frac{7}{16}$

$2\frac{1}{8}$

6

FIXTURE BLOCK
SCALE: 6"=1'–0
CRS – 2 REQD

SHOP NOTES AND DIMENSIONS	DRAWN BY	FILE NO.	DRAWING
VERTICAL LETTERING			**A-2**

| | DRAWN BY | | FILE NO. | DRAWING |

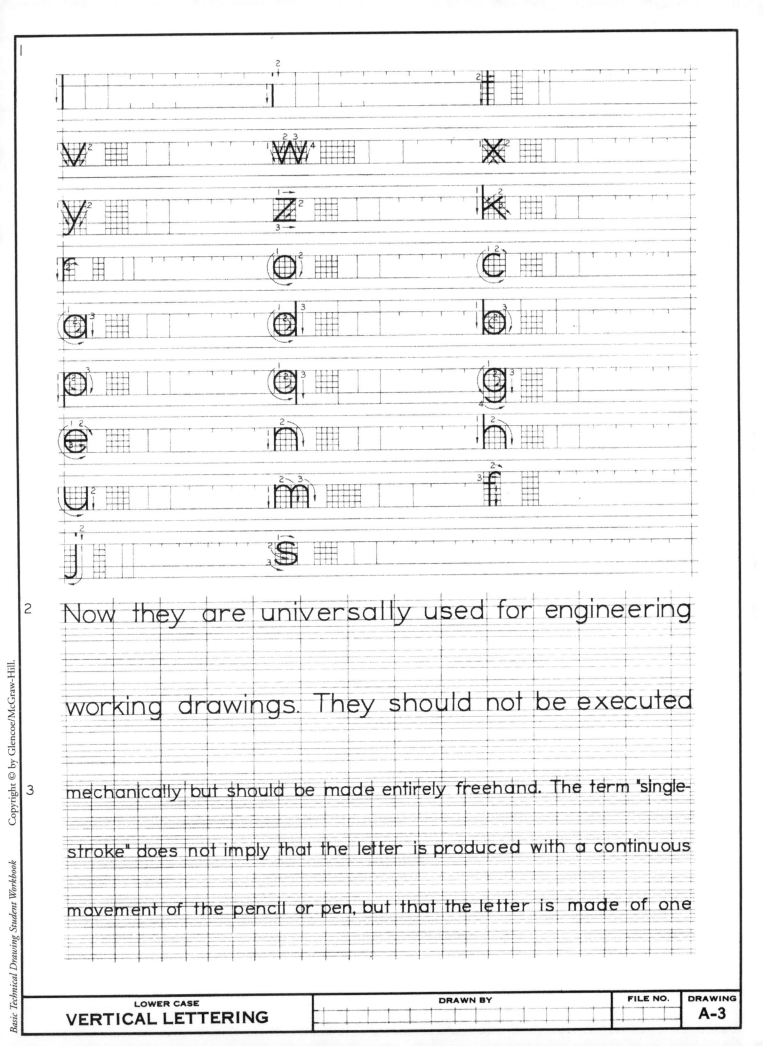

Now they are universally used for engineering

working drawings. They should not be executed

mechanically but should be made entirely freehand. The term "single-

stroke" does not imply that the letter is produced with a continuous

movement of the pencil or pen, but that the letter is made of one

LOWER CASE
VERTICAL LETTERING

DRAWN BY

FILE NO.

DRAWING
A-3

	DRAWN BY	FILE NO.	DRAWING

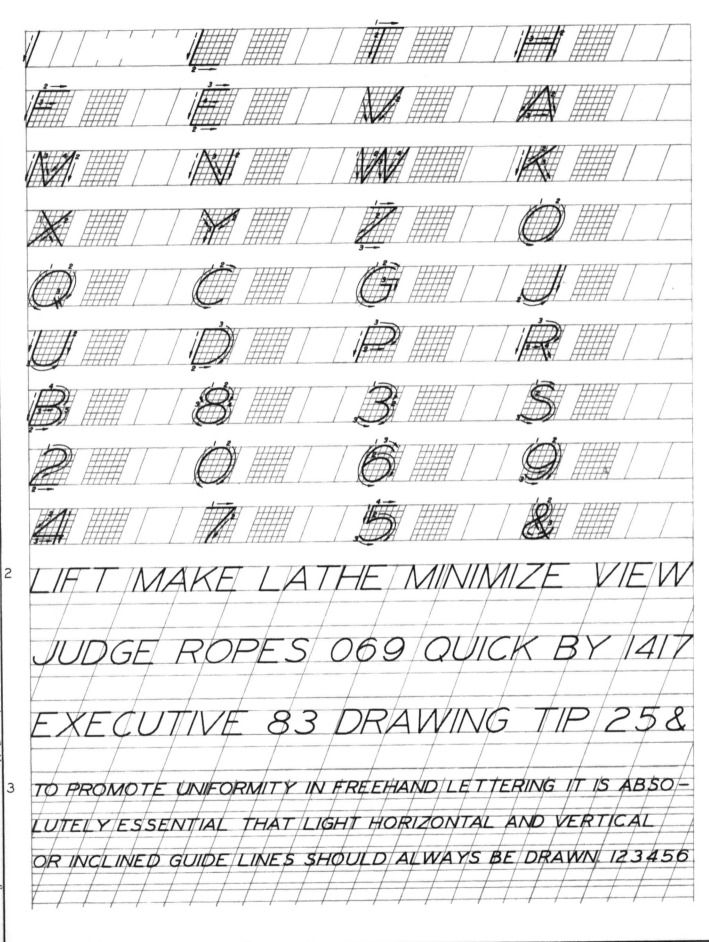

LIFT MAKE LATHE MINIMIZE VIEW

JUDGE ROPES 069 QUICK BY 1417

EXECUTIVE 83 DRAWING TIP 25&

TO PROMOTE UNIFORMITY IN FREEHAND LETTERING IT IS ABSO-

LUTELY ESSENTIAL THAT LIGHT HORIZONTAL AND VERTICAL

OR INCLINED GUIDE LINES SHOULD ALWAYS BE DRAWN 123456

CAPITALS AND NUMERALS
INCLINED LETTERING

DRAWN BY

FILE NO.

DRAWING
B-1

| | | DRAWN BY | FILE NO. | DRAWING |

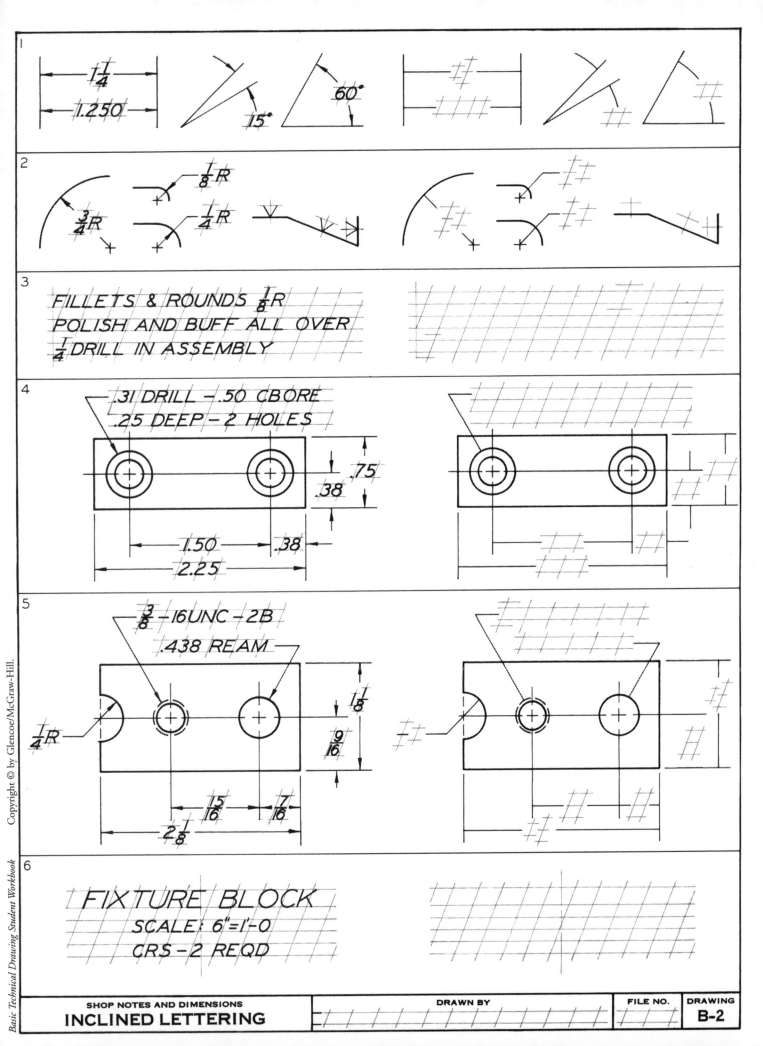

1

$1\frac{1}{4}$

1.250

15°

60°

2

$\frac{3}{4}R$

$\frac{1}{8}R$

$\frac{1}{4}R$

3

FILLETS & ROUNDS $\frac{1}{8}R$
POLISH AND BUFF ALL OVER
$\frac{1}{4}$ DRILL IN ASSEMBLY

4

.31 DRILL – .50 CBORE
.25 DEEP – 2 HOLES

.75

.38

1.50

.38

2.25

5

$\frac{3}{8} - 16UNC - 2B$

.438 REAM

$\frac{1}{4}R$

$1\frac{1}{8}$

$\frac{9}{16}$

$\frac{15}{16}$

$\frac{7}{16}$

$2\frac{1}{8}$

6

FIXTURE BLOCK
SCALE: 6"=1'-0
CRS–2 REQD

SHOP NOTES AND DIMENSIONS **INCLINED LETTERING**	DRAWN BY	FILE NO.	DRAWING **B-2**

Basic Technical Drawing Student Workbook

		DRAWN BY		FILE NO.	DRAWING

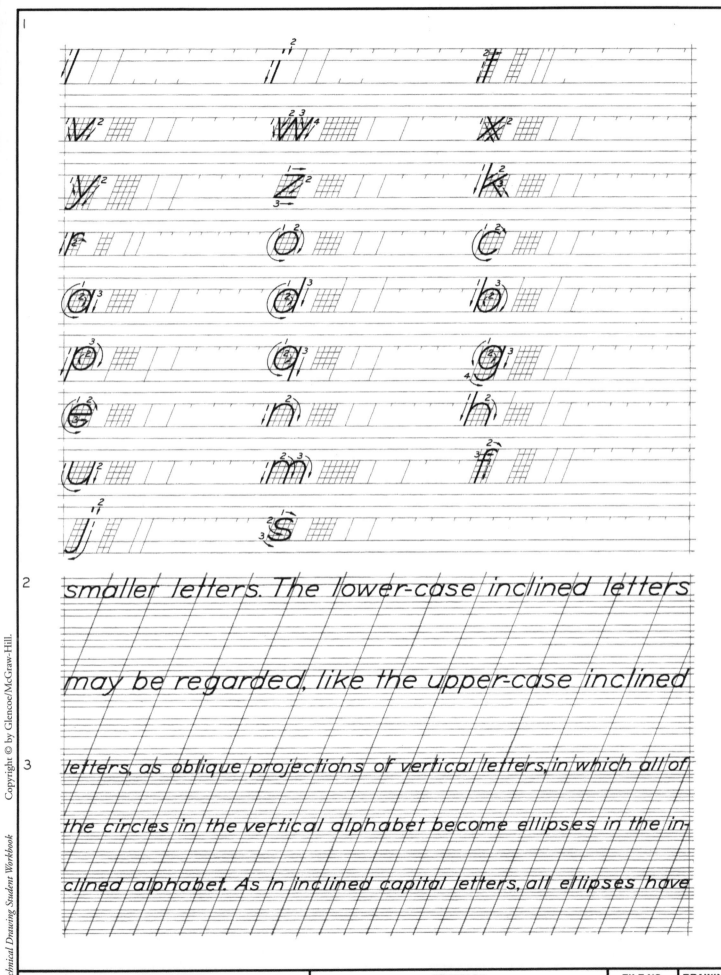

smaller letters. The lower-case inclined letters

may be regarded, like the upper-case inclined

letters, as oblique projections of vertical letters, in which all of

the circles in the vertical alphabet become ellipses in the in-

clined alphabet. As in inclined capital letters, all ellipses have

| LOWER CASE
INCLINED LETTERING | DRAWN BY | FILE NO. | DRAWING
B-3 |

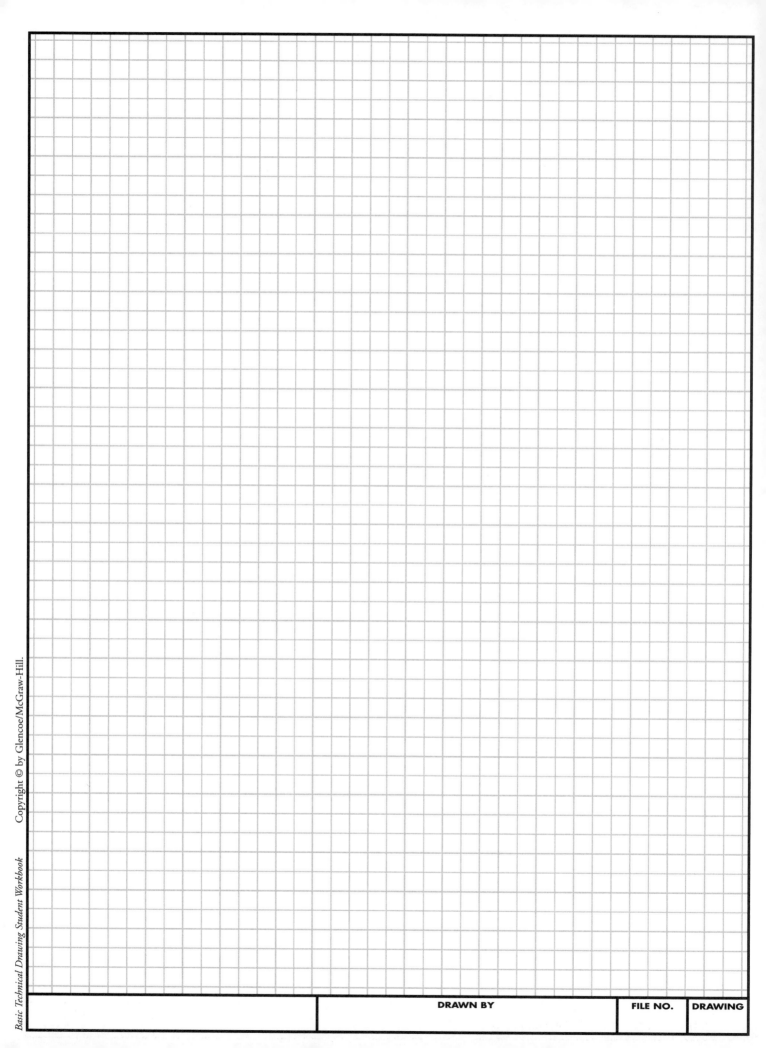

Basic Technical Drawing Student Workbook

DRAWN BY

FILE NO.

DRAWING

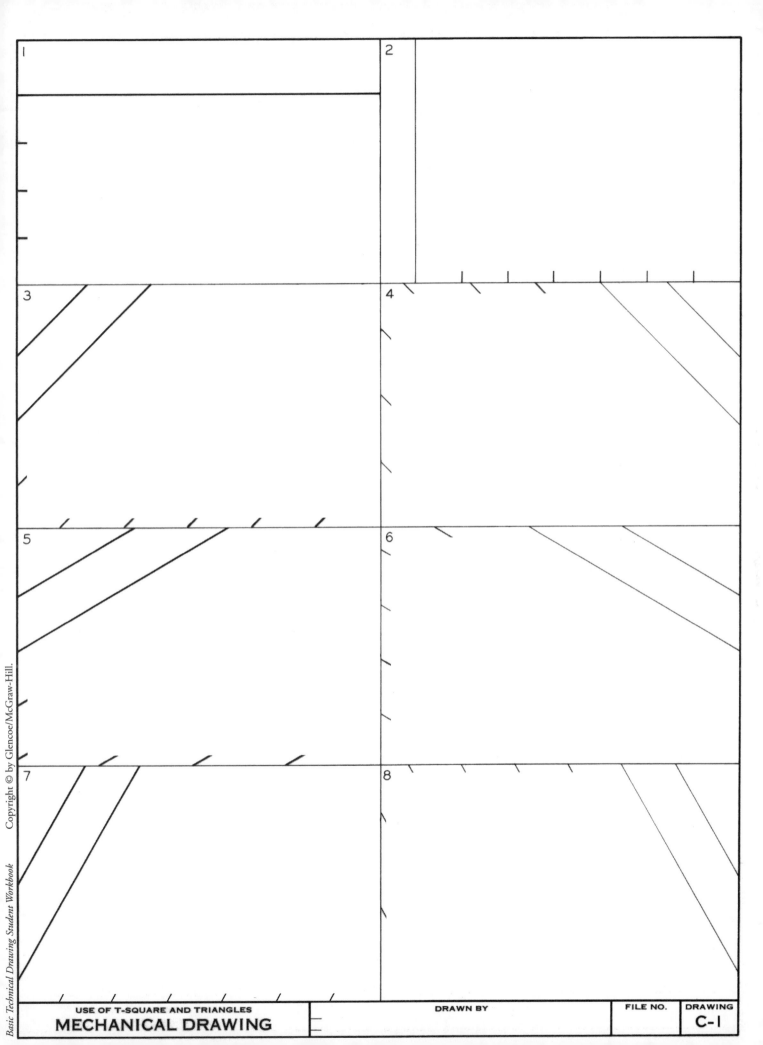

1

2

3

4

5

6

7

8

USE OF T-SQUARE AND TRIANGLES

MECHANICAL DRAWING

DRAWN BY

FILE NO.

DRAWING

C-1

| | | DRAWN BY | FILE NO. | DRAWING |

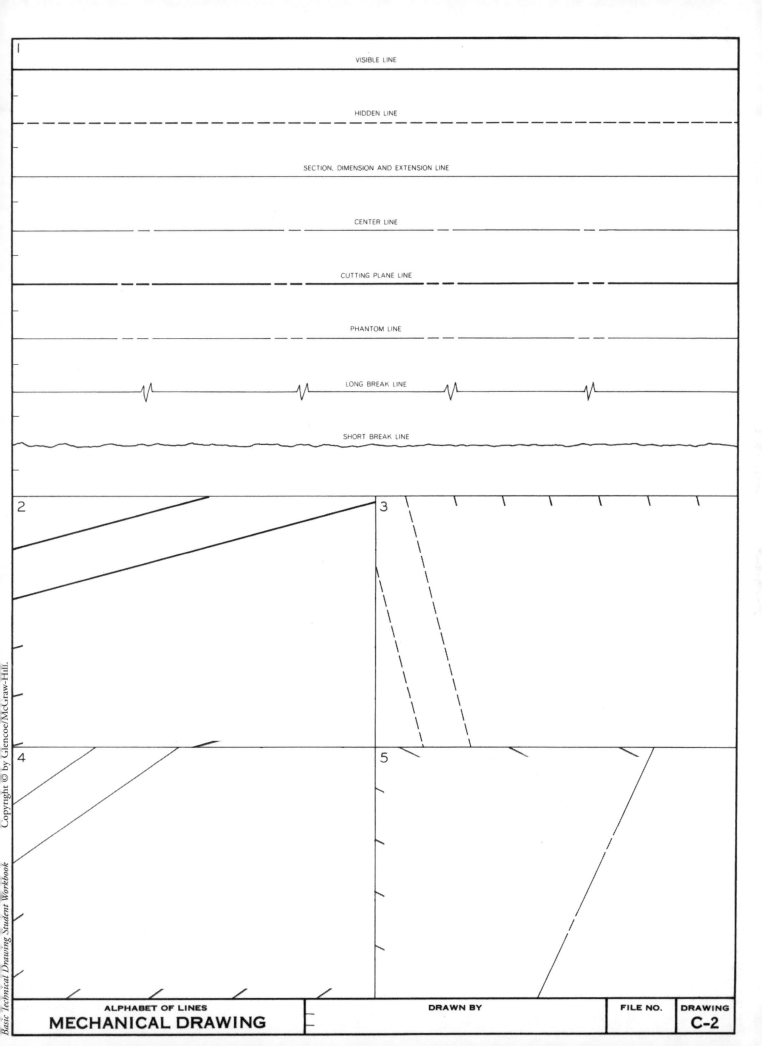

1

VISIBLE LINE

HIDDEN LINE

SECTION, DIMENSION AND EXTENSION LINE

CENTER LINE

CUTTING PLANE LINE

PHANTOM LINE

LONG BREAK LINE

SHORT BREAK LINE

2

3

4

5

| ALPHABET OF LINES | DRAWN BY | FILE NO. | DRAWING |
| MECHANICAL DRAWING | | | C-2 |

Basic Technical Drawing Student Workbook

	DRAWN BY	FILE NO.	DRAWING

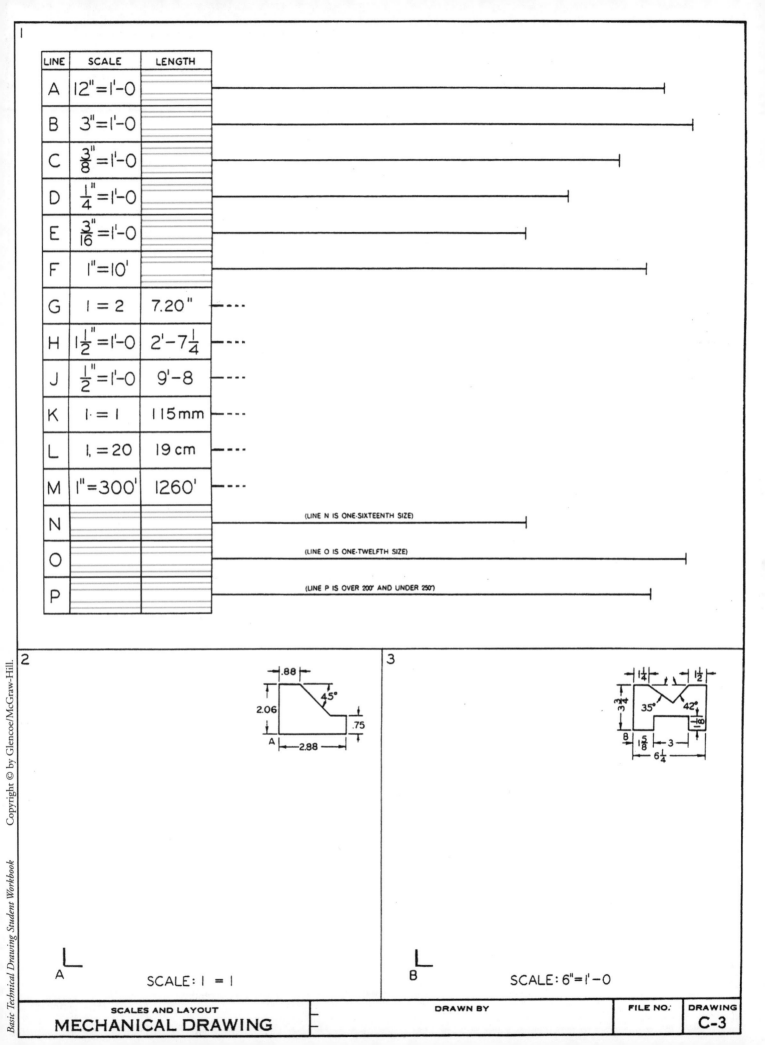

LINE	SCALE	LENGTH
A	12"=1'-0	
B	3"=1'-0	
C	$\frac{3}{8}$"=1'-0	
D	$\frac{1}{4}$"=1'-0	
E	$\frac{3}{16}$"=1'-0	
F	1"=10'	
G	1 = 2	7.20"
H	$1\frac{1}{2}$"=1'-0	$2'-7\frac{1}{4}$
J	$\frac{1}{2}$"=1'-0	9'-8
K	1 = 1	115mm
L	1 = 20	19 cm
M	1"=300'	1260'
N		
O		
P		

(LINE N IS ONE-SIXTEENTH SIZE)

(LINE O IS ONE-TWELFTH SIZE)

(LINE P IS OVER 200' AND UNDER 250')

2

.88
45°
2.06
.75
A
2.88

3

$1\frac{1}{4}$ $1\frac{1}{2}$
$3\frac{3}{4}$
35° 42°
$\frac{1}{8}$
B $1\frac{5}{8}$ 3
$6\frac{1}{4}$

A
SCALE: 1 = 1

B
SCALE: 6"=1'-0

SCALES AND LAYOUT	DRAWN BY	FILE NO:	DRAWING
MECHANICAL DRAWING			**C-3**

		DRAWN BY		FILE NO.	DRAWING

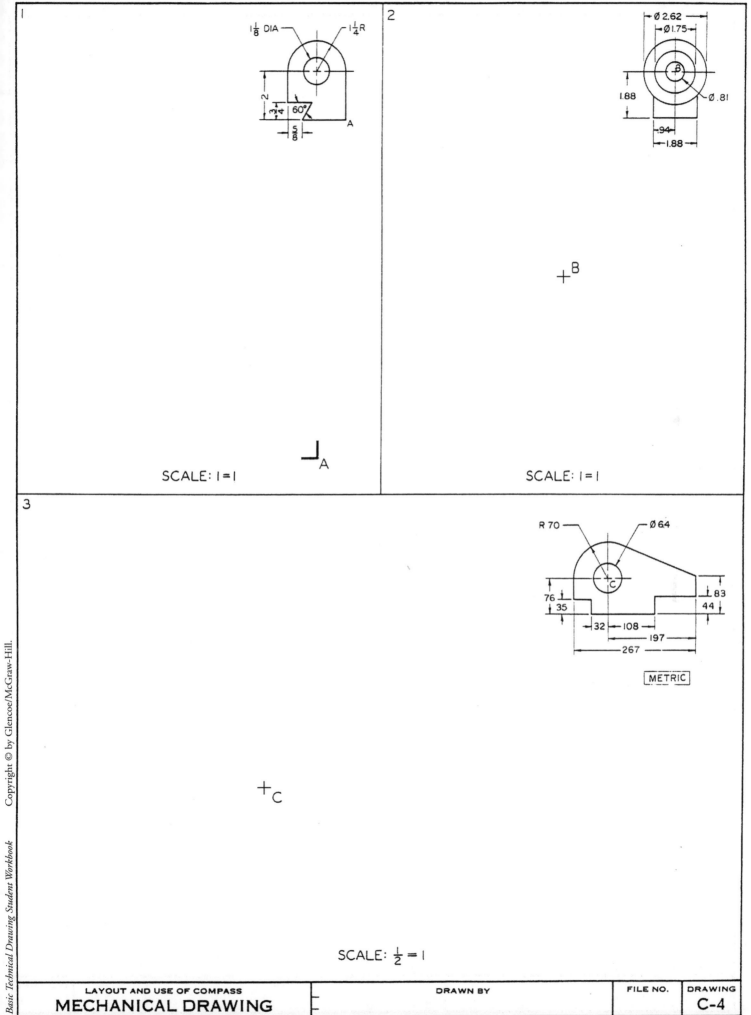

1

$1\frac{1}{8}$ DIA $1\frac{1}{4}$R

2

$\frac{3}{4}$ 60°

$\frac{5}{8}$

A

SCALE: 1=1

2

∅ 2.62

∅ 1.75

B

∅ .81

1.88

94

1.88

+B

SCALE: 1=1

3

R 70 ∅ 64

C

76

35

83

44

32 108

197

267

METRIC

+C

SCALE: $\frac{1}{2}$ = 1

| LAYOUT AND USE OF COMPASS | DRAWN BY | FILE NO. | DRAWING |
| MECHANICAL DRAWING | | | C-4 |

	DRAWN BY	FILE NO.	DRAWING

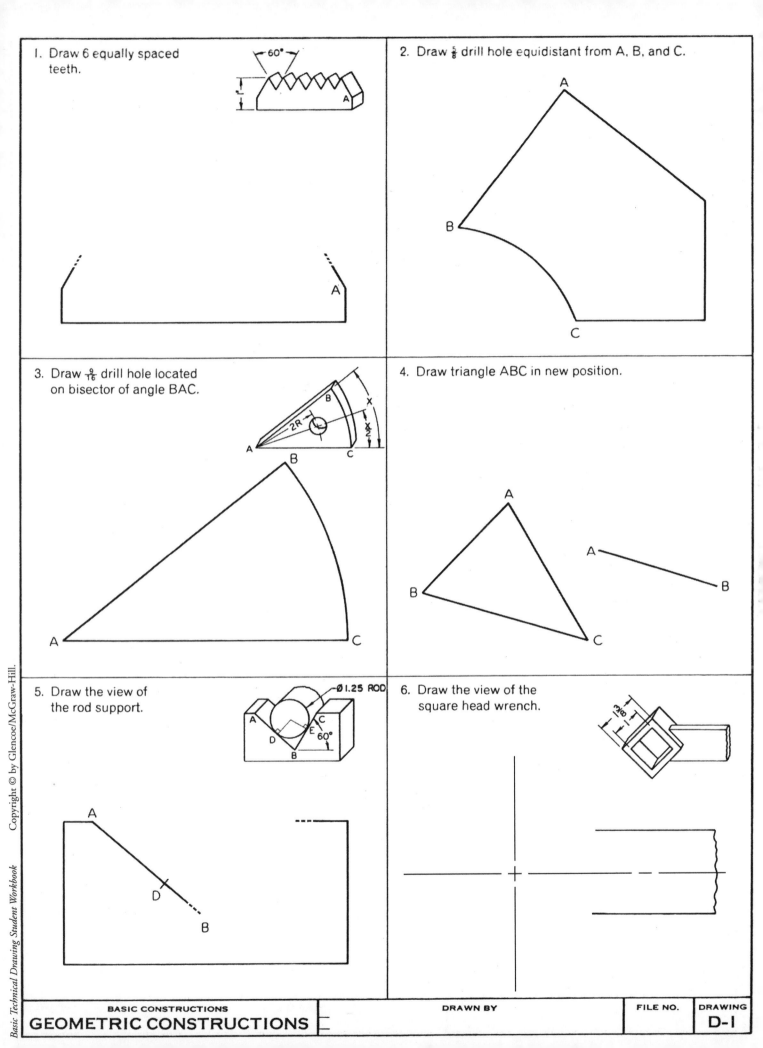

1. Draw 6 equally spaced teeth.

2. Draw $\frac{5}{8}$ drill hole equidistant from A, B, and C.

3. Draw $\frac{9}{16}$ drill hole located on bisector of angle BAC.

4. Draw triangle ABC in new position.

5. Draw the view of the rod support.

6. Draw the view of the square head wrench.

BASIC CONSTRUCTIONS
GEOMETRIC CONSTRUCTIONS

DRAWN BY

FILE NO.

DRAWING
D-1

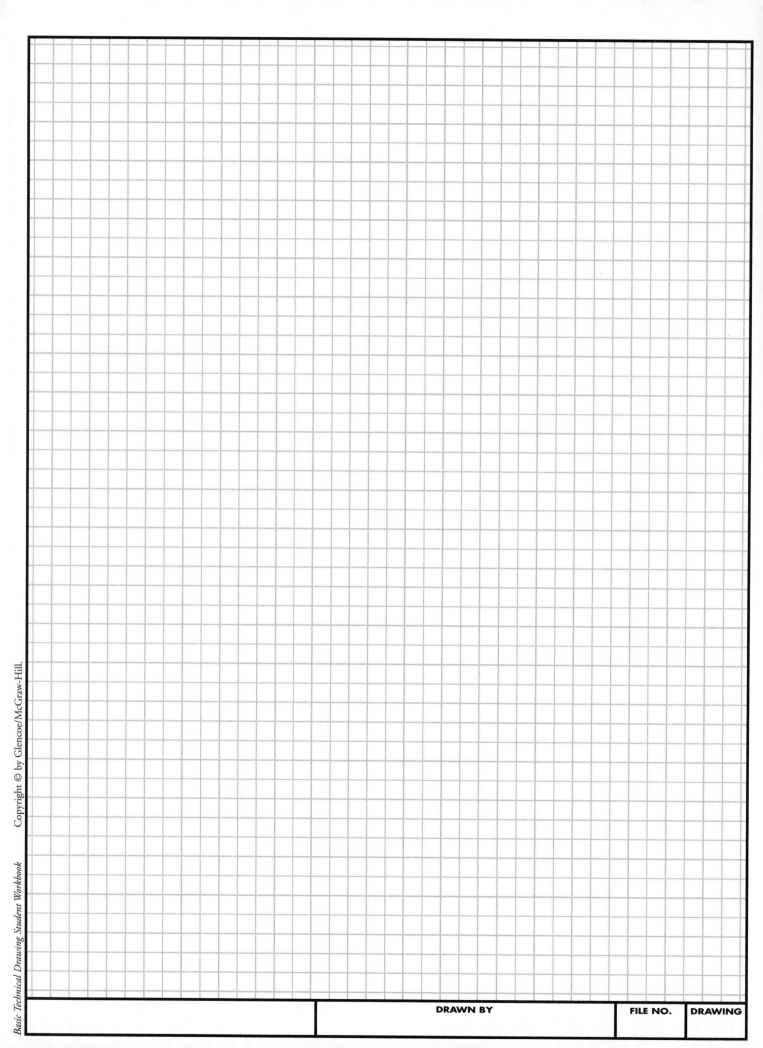

DRAWN BY

FILE NO.

DRAWING

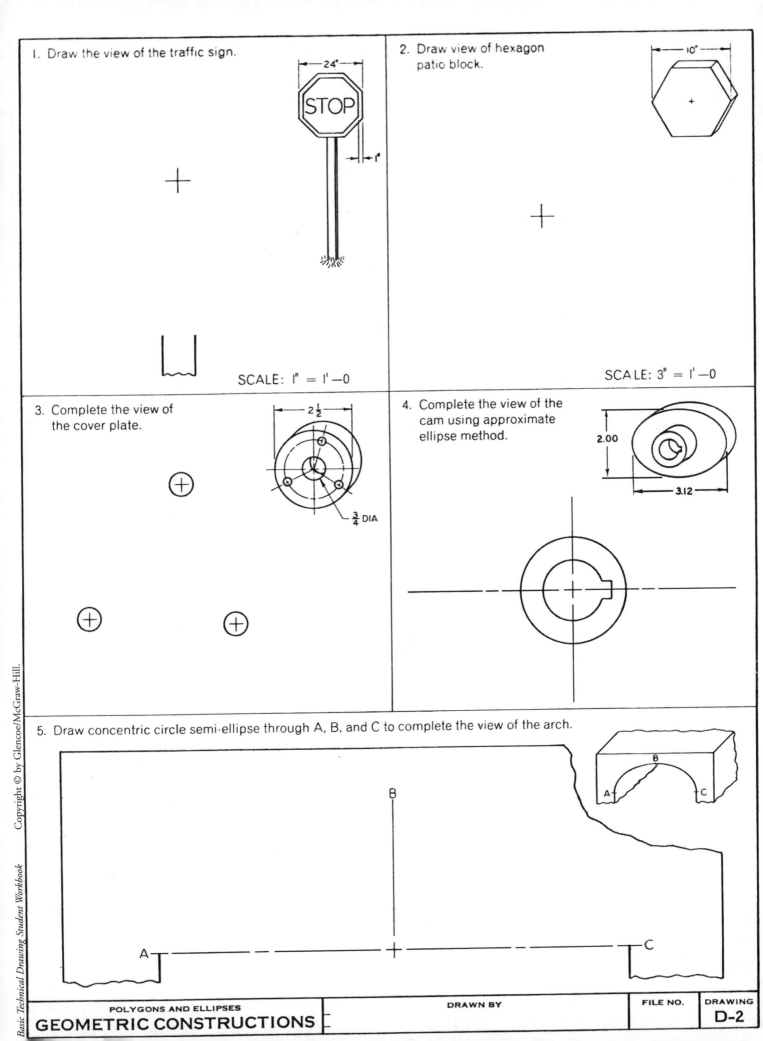

1. Draw the view of the traffic sign.

24"

STOP

1"

SCALE: 1" = 1'—0

2. Draw view of hexagon patio block.

10"

SCALE: 3" = 1'—0

3. Complete the view of the cover plate.

2½

¾ DIA

4. Complete the view of the cam using approximate ellipse method.

2.00

3.12

5. Draw concentric circle semi-ellipse through A, B, and C to complete the view of the arch.

A   B   C

A   B   C

POLYGONS AND ELLIPSES

GEOMETRIC CONSTRUCTIONS

DRAWN BY

FILE NO.

DRAWING

D-2

Basic Technical Drawing Student Workbook

		DRAWN BY	FILE NO.	DRAWING

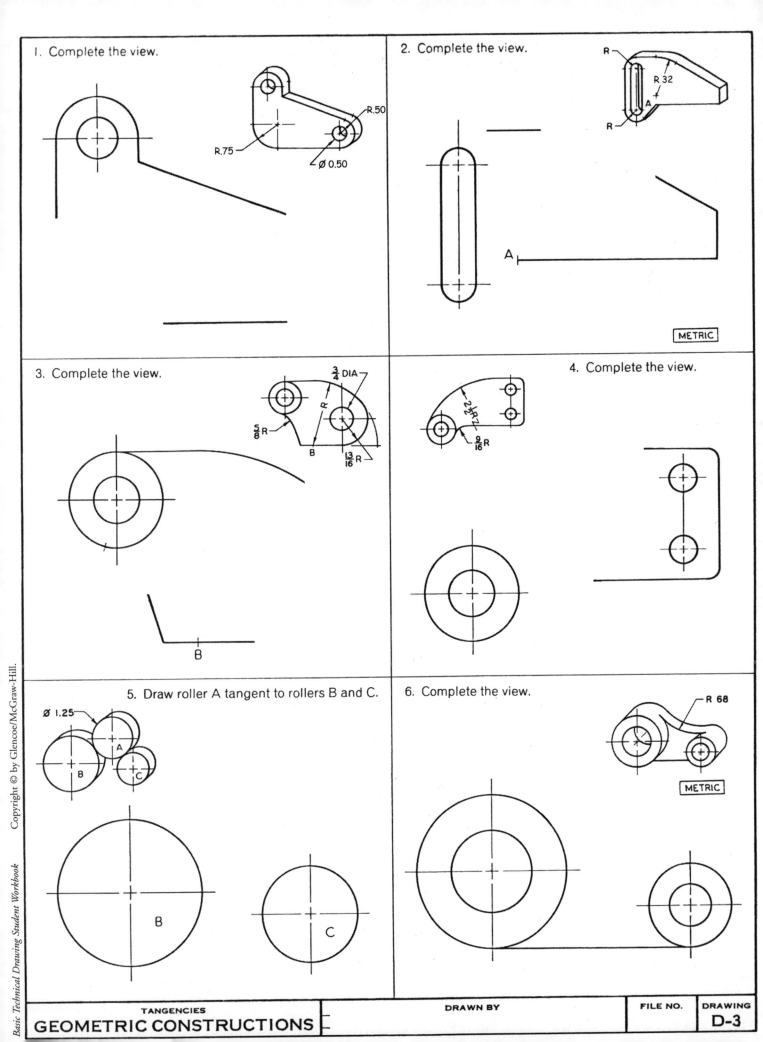

1. Complete the view.

R.50
R.75
Ø 0.50

2. Complete the view.

R
R 32
A
R
A
METRIC

3. Complete the view.

3/4 DIA
R
5/8 R
B
13/16 R

4. Complete the view.

2 1/2 R
9/16 R

5. Draw roller A tangent to rollers B and C.

Ø 1.25
A
B
C
B
C

6. Complete the view.

R 68
METRIC

TANGENCIES

GEOMETRIC CONSTRUCTIONS

DRAWN BY

FILE NO.

DRAWING

D-3

		DRAWN BY		FILE NO.	DRAWING

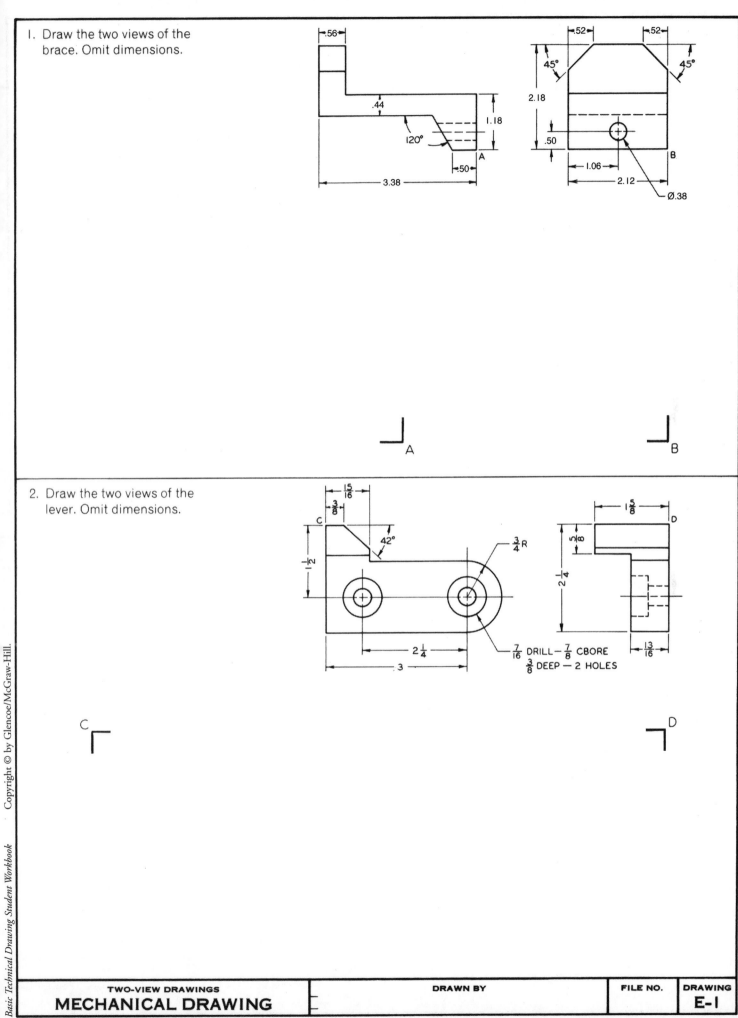

1. Draw the two views of the brace. Omit dimensions.

.56

.44

120°

1.18

.50

A

3.38

.52   .52

45°   45°

2.18

.50

1.06

2.12

B

Ø.38

A

B

2. Draw the two views of the lever. Omit dimensions.

$\frac{15}{16}$

$\frac{3}{8}$

C

42°

$\frac{3}{4}$R

$1\frac{1}{2}$

$1\frac{5}{8}$

D

$\frac{5}{8}$

$2\frac{1}{4}$

$2\frac{1}{4}$

3

$\frac{13}{16}$

$\frac{7}{16}$ DRILL – $\frac{7}{8}$ CBORE
$\frac{3}{8}$ DEEP – 2 HOLES

C

D

TWO-VIEW DRAWINGS	DRAWN BY	FILE NO.	DRAWING
MECHANICAL DRAWING			E-1

| | DRAWN BY | FILE NO. | DRAWING |

Draw the three views of
the bracket. Omit
dimensions unless
assigned.
SCALE: HALF SIZE

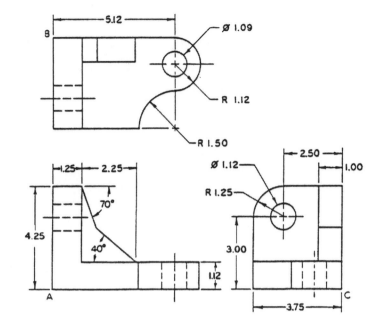

B ⌐

L
A

⌐ B

⌐ C

THREE-VIEW DRAWING	DRAWN BY	FILE NO.	DRAWING
MECHANICAL DRAWING			E-2

Basic Technical Drawing Student Workbook

| | | DRAWN BY | FILE NO. | DRAWING |

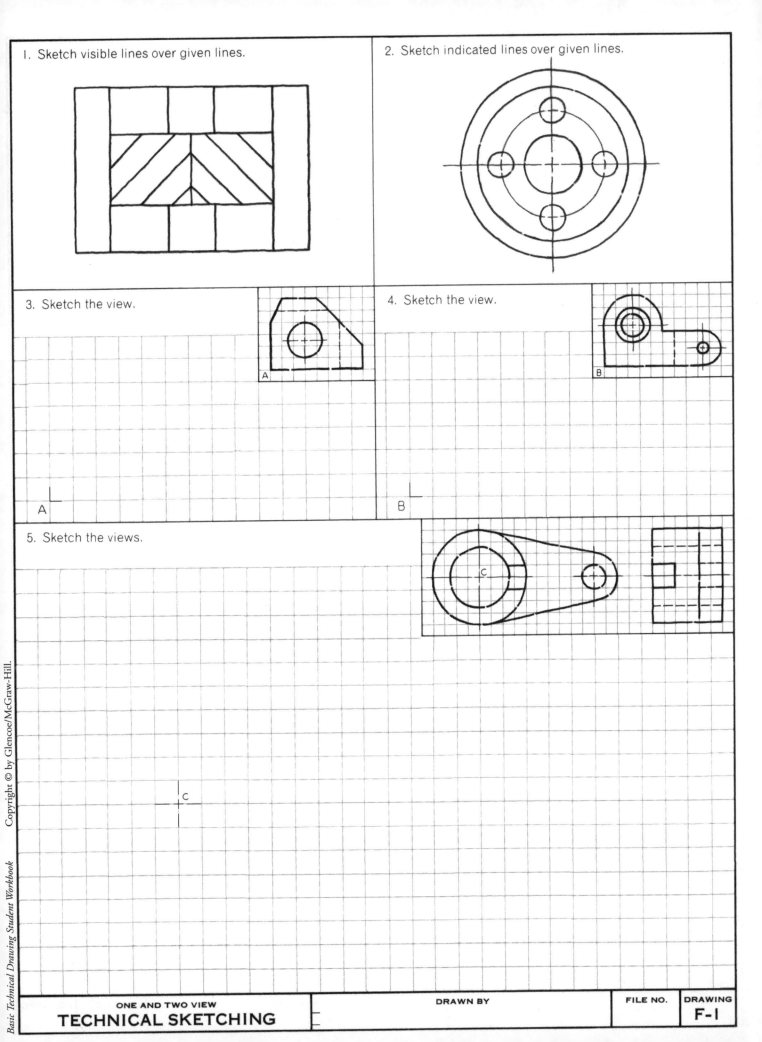

1. Sketch visible lines over given lines.

2. Sketch indicated lines over given lines.

3. Sketch the view.

A

4. Sketch the view.

B

5. Sketch the views.

C

ONE AND TWO VIEW
TECHNICAL SKETCHING

DRAWN BY

FILE NO.

DRAWING
F-1

		DRAWN BY		FILE NO.	DRAWING

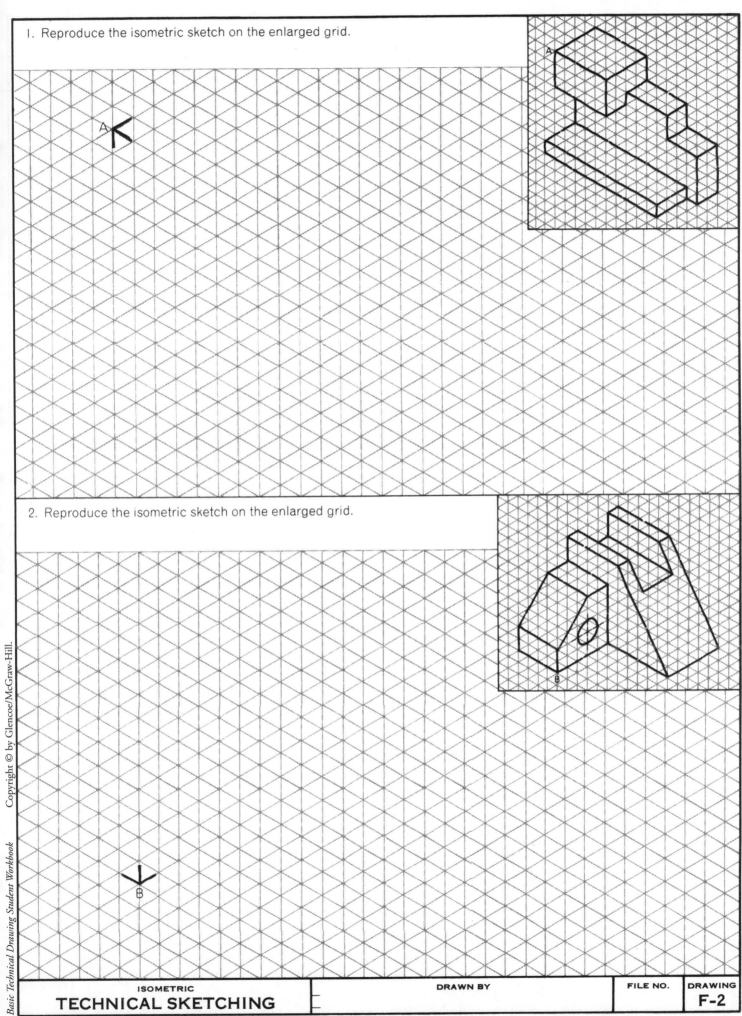

1. Reproduce the isometric sketch on the enlarged grid.

A

2. Reproduce the isometric sketch on the enlarged grid.

B

ISOMETRIC
TECHNICAL SKETCHING

DRAWN BY

FILE NO.

DRAWING
F-2

	DRAWN BY		FILE NO.	DRAWING

1. Reproduce the oblique sketch on the enlarged grid.

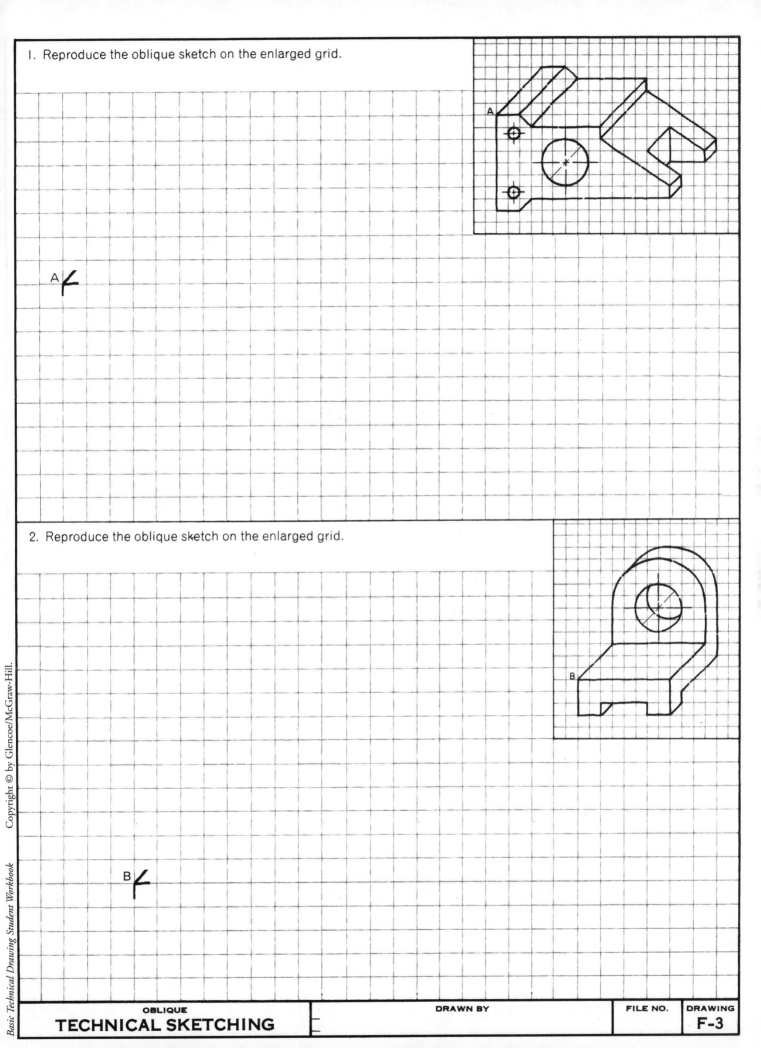

A

2. Reproduce the oblique sketch on the enlarged grid.

B

OBLIQUE **TECHNICAL SKETCHING**		DRAWN BY	FILE NO.	DRAWING **F-3**

Basic Technical Drawing Student Workbook

		DRAWN BY	FILE NO.	DRAWING

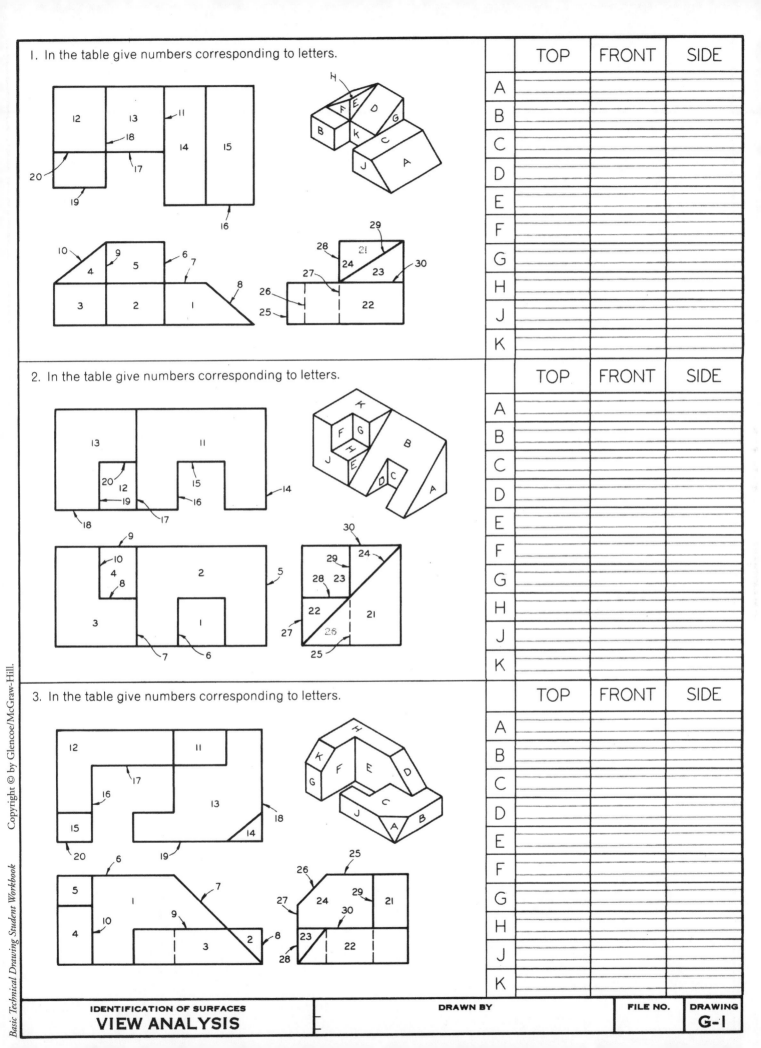

1. In the table give numbers corresponding to letters.

	TOP	FRONT	SIDE
A			
B			
C			
D			
E			
F			
G			
H			
J			
K			

2. In the table give numbers corresponding to letters.

	TOP	FRONT	SIDE
A			
B			
C			
D			
E			
F			
G			
H			
J			
K			

3. In the table give numbers corresponding to letters.

	TOP	FRONT	SIDE
A			
B			
C			
D			
E			
F			
G			
H			
J			
K			

IDENTIFICATION OF SURFACES
VIEW ANALYSIS

DRAWN BY

FILE NO.

DRAWING
G-1

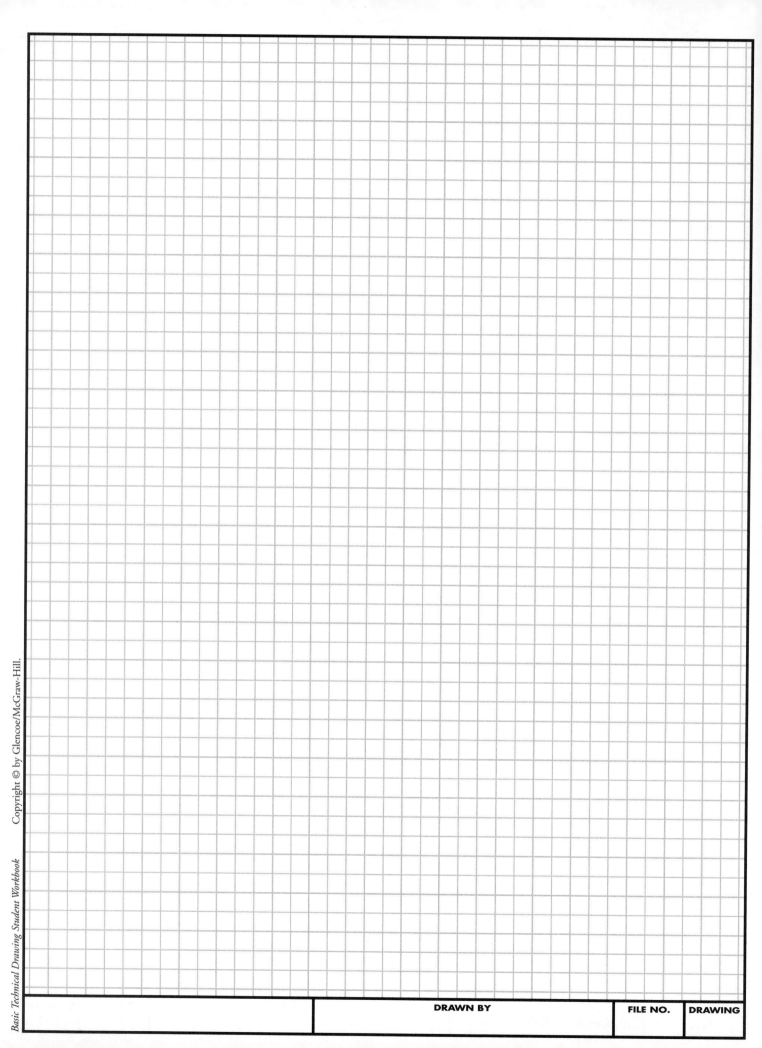

Basic Technical Drawing Student Workbook

DRAWN BY

FILE NO.

DRAWING

SUPPORT

Make a full size 6-view freehand sketch of the object, with the views in the standard arrangement. Label all views: front, top, etc. Space the views two squares apart. Show all hidden lines and center lines.

FRONT

A ⌐

FRONT

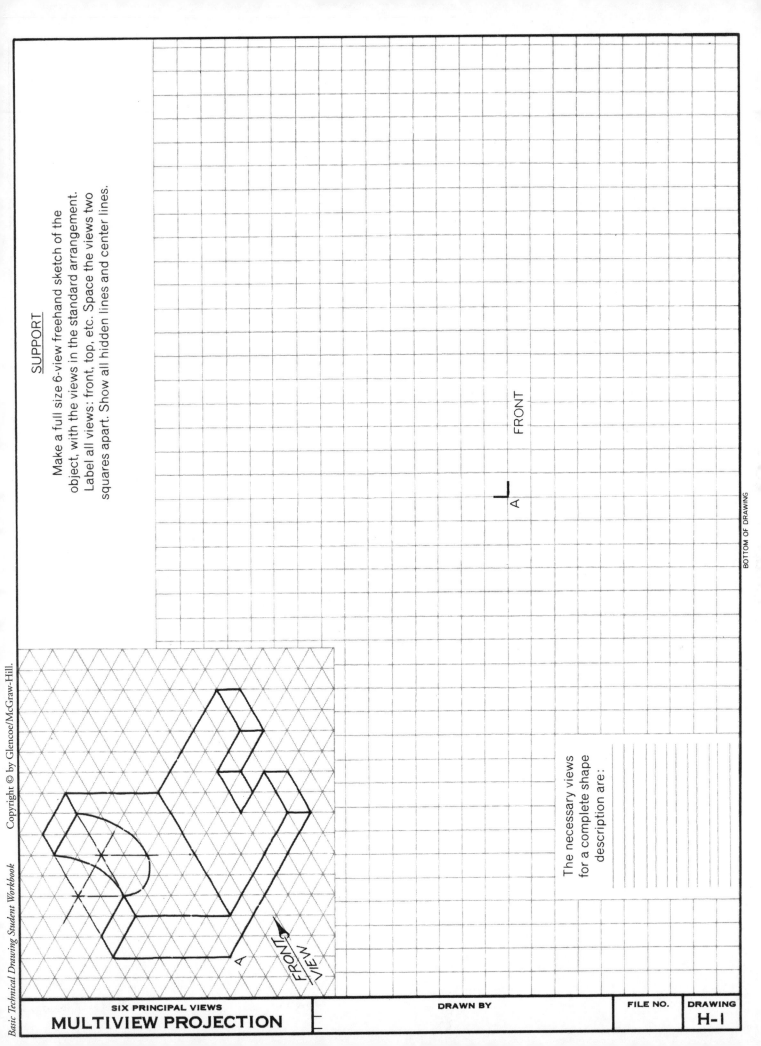

A

FRONT VIEW

The necessary views for a complete shape description are:

BOTTOM OF DRAWING

SIX PRINCIPAL VIEWS	DRAWN BY	FILE NO.	DRAWING
MULTIVIEW PROJECTION			**H-1**

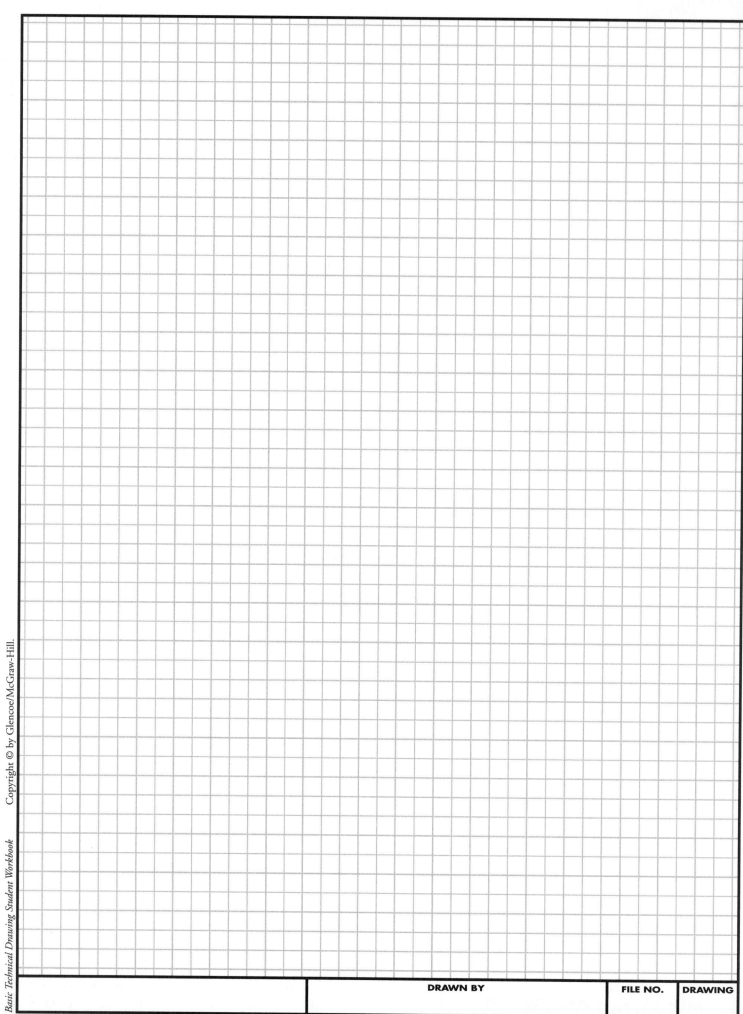

Basic Technical Drawing Student Workbook

DRAWN BY

FILE NO.

DRAWING

TOOL GUIDE

Make a full-size 6-view freehand sketch of the object, with the views in the standard arrangement. Label all views: front, top, etc. Space the views two squares apart. Show all hidden lines and center lines.

FRONT

A

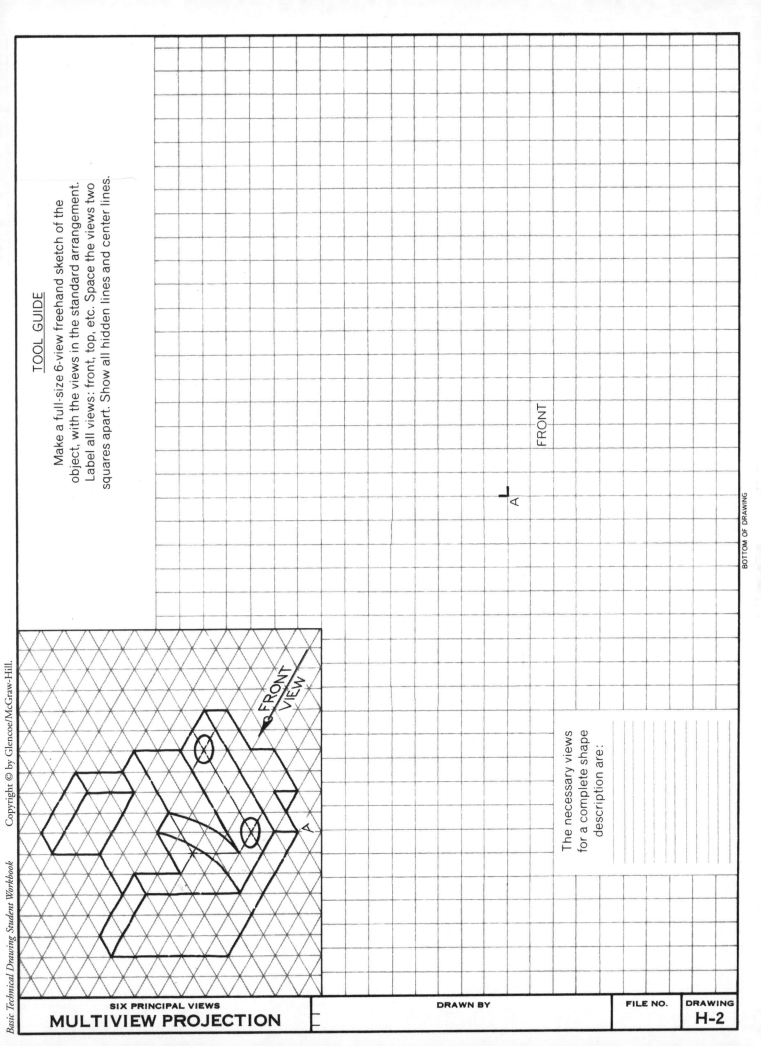

FRONT VIEW

BOTTOM OF DRAWING

The necessary views for a complete shape description are:

SIX PRINCIPAL VIEWS	DRAWN BY	FILE NO.	DRAWING
MULTIVIEW PROJECTION			**H-2**

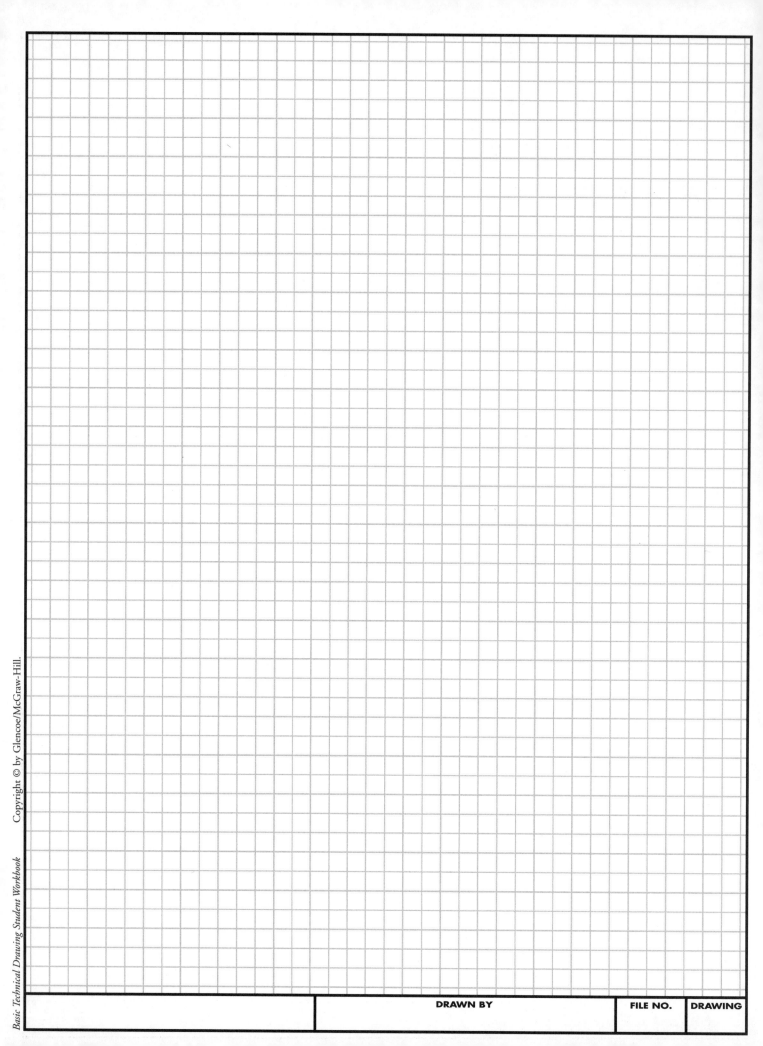

Basic Technical Drawing Student Workbook

		DRAWN BY		FILE NO.	DRAWING

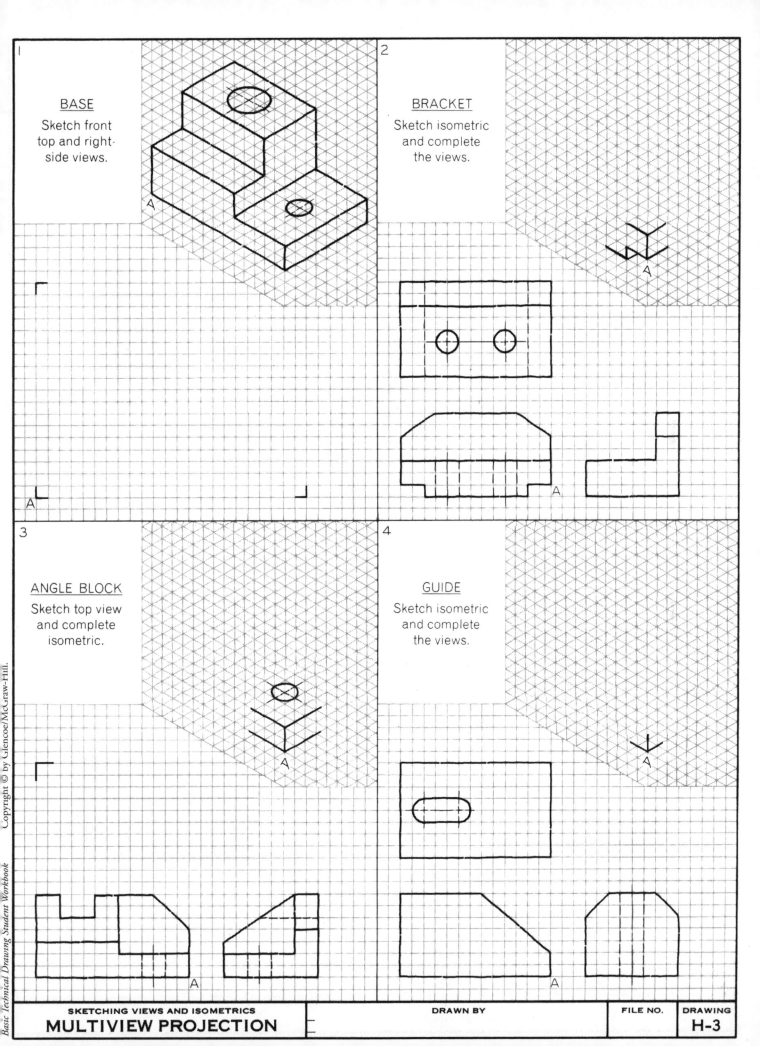

1

<u>BASE</u>

Sketch front top and right-side views.

A

2

<u>BRACKET</u>

Sketch isometric and complete the views.

A

A

3

<u>ANGLE BLOCK</u>

Sketch top view and complete isometric.

A

A

4

<u>GUIDE</u>

Sketch isometric and complete the views.

A

A

Basic Technical Drawing Student Workbook   Copyright © by Glencoe/McGraw-Hill.

SKETCHING VIEWS AND ISOMETRICS
MULTIVIEW PROJECTION

DRAWN BY

FILE NO.

DRAWING
H-3

| | | DRAWN BY | FILE NO. | DRAWING |

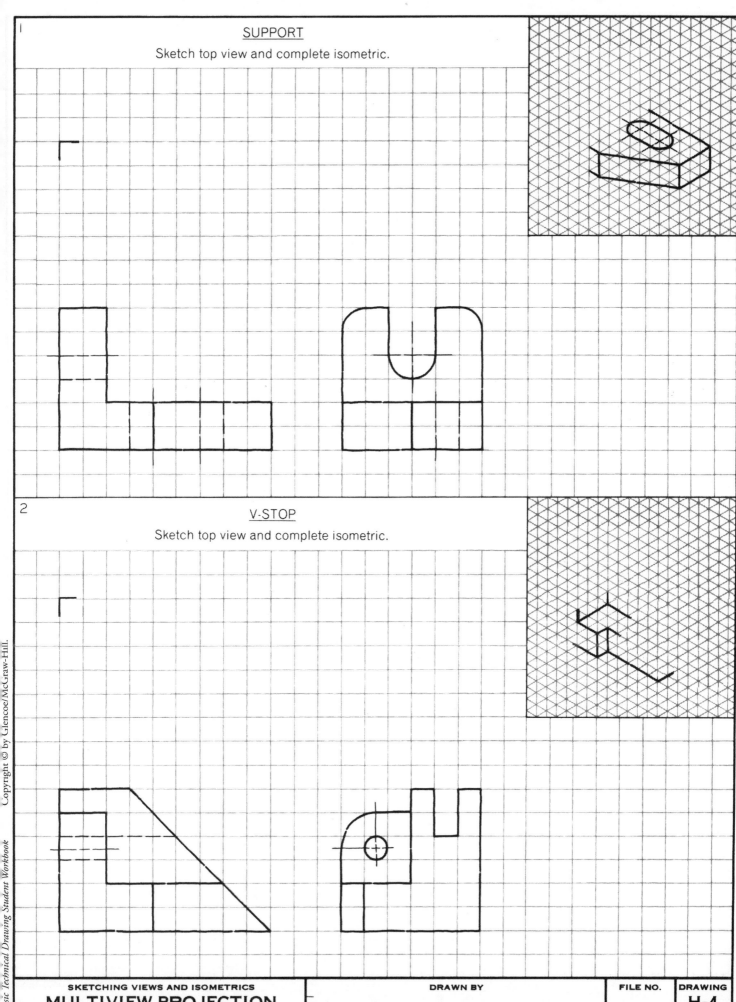

SUPPORT

Sketch top view and complete isometric.

V-STOP

Sketch top view and complete isometric.

	DRAWN BY	FILE NO.	DRAWING

MISSING LINES
MULTIVIEW PROJECTION

DRAWN BY

FILE NO.

DRAWING
H-5

| | | DRAWN BY | | FILE NO. | DRAWING |

MISSING LINES

MULTIVIEW PROJECTION

DRAWN BY

FILE NO.

DRAWING

H-6

Basic Technical Drawing Student Workbook

		DRAWN BY		FILE NO.	DRAWING

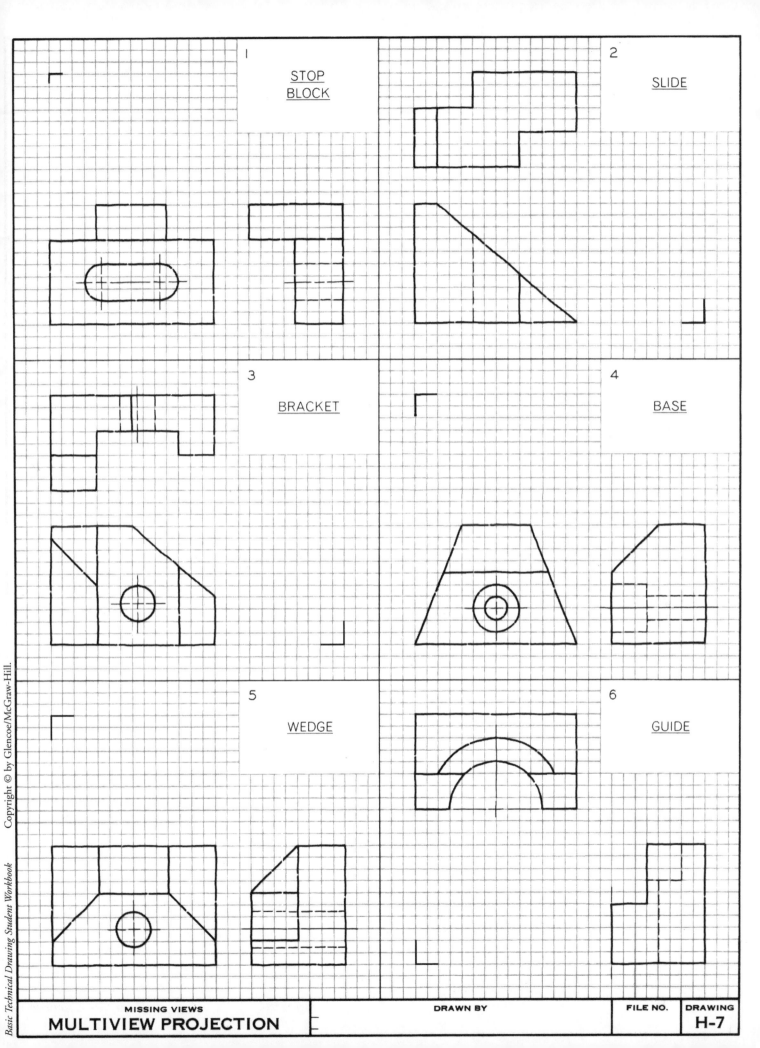

1 STOP BLOCK

2 SLIDE

3 BRACKET

4 BASE

5 WEDGE

6 GUIDE

MISSING VIEWS

MULTIVIEW PROJECTION

DRAWN BY

FILE NO.

DRAWING

H-7

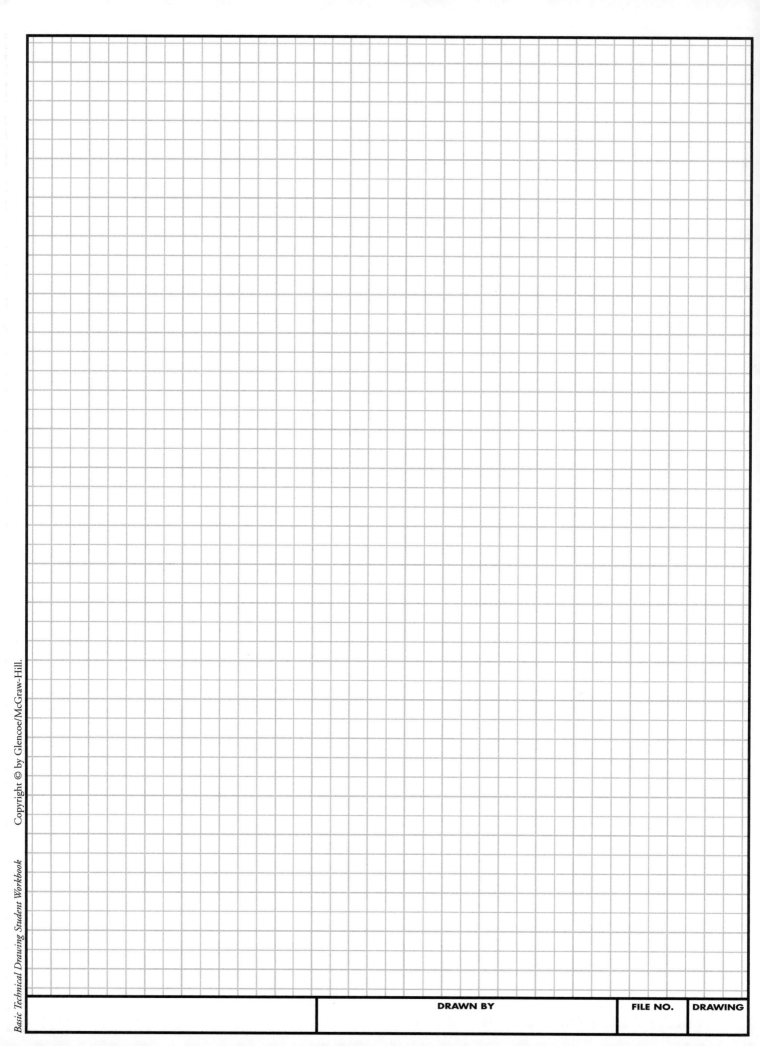

Basic Technical Drawing Student Workbook

		DRAWN BY		FILE NO.	DRAWING

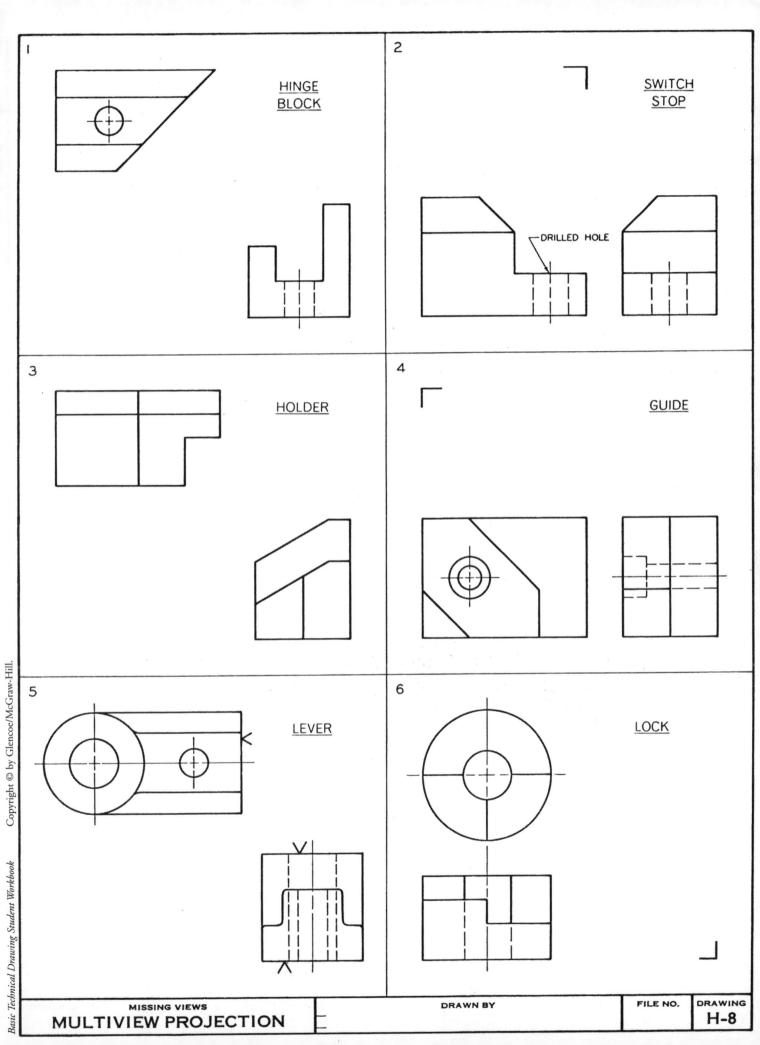

1

HINGE
BLOCK

2

SWITCH
STOP

DRILLED HOLE

3

HOLDER

4

GUIDE

5

LEVER

6

LOCK

MISSING VIEWS

MULTIVIEW PROJECTION

DRAWN BY

FILE NO.

DRAWING

H-8

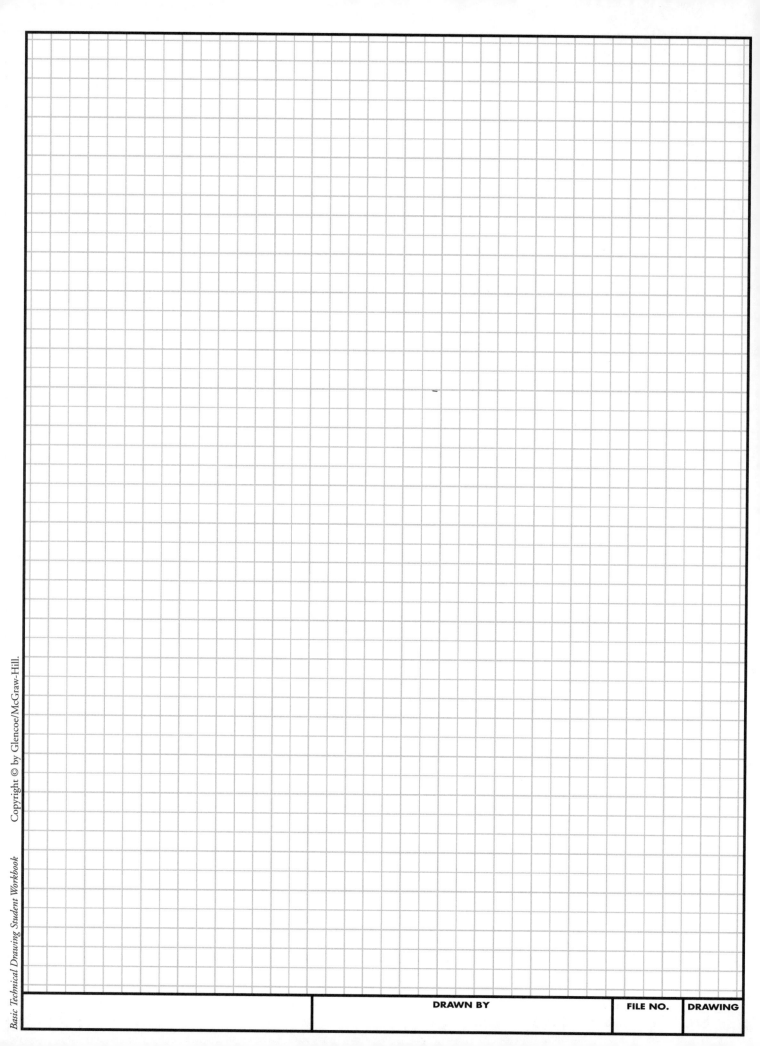

	DRAWN BY	FILE NO.	DRAWING

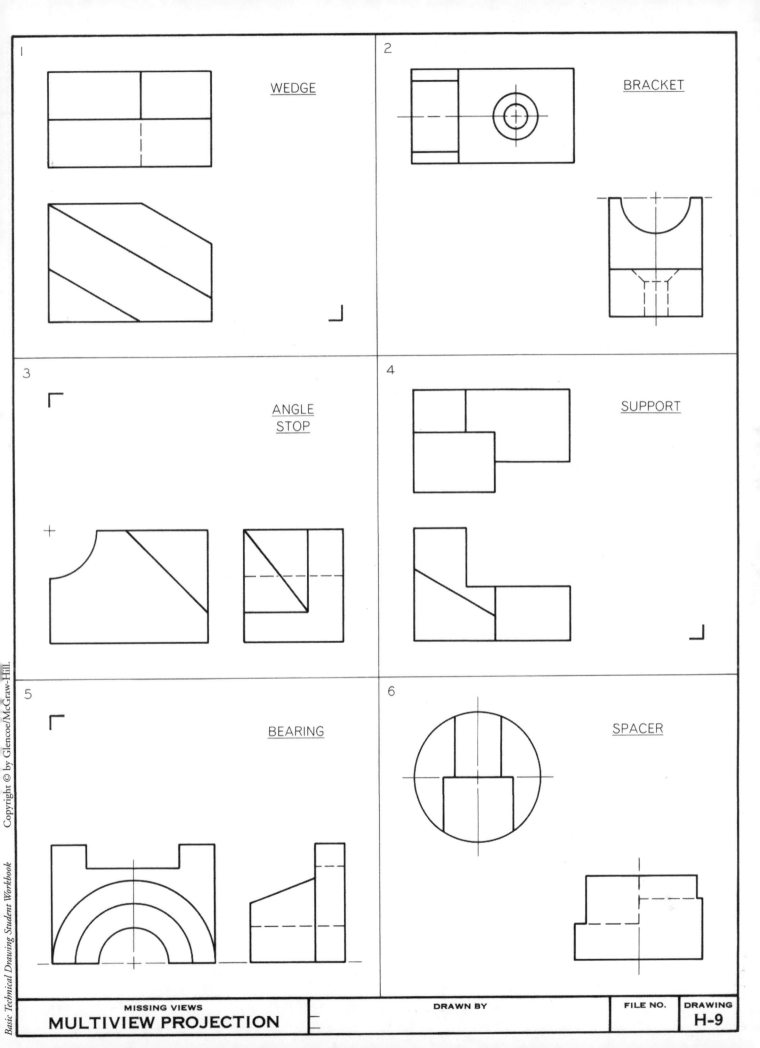

1　WEDGE

2　BRACKET

3　ANGLE STOP

4　SUPPORT

5　BEARING

6　SPACER

MISSING VIEWS
MULTIVIEW PROJECTION

DRAWN BY

FILE NO.

DRAWING
H-9

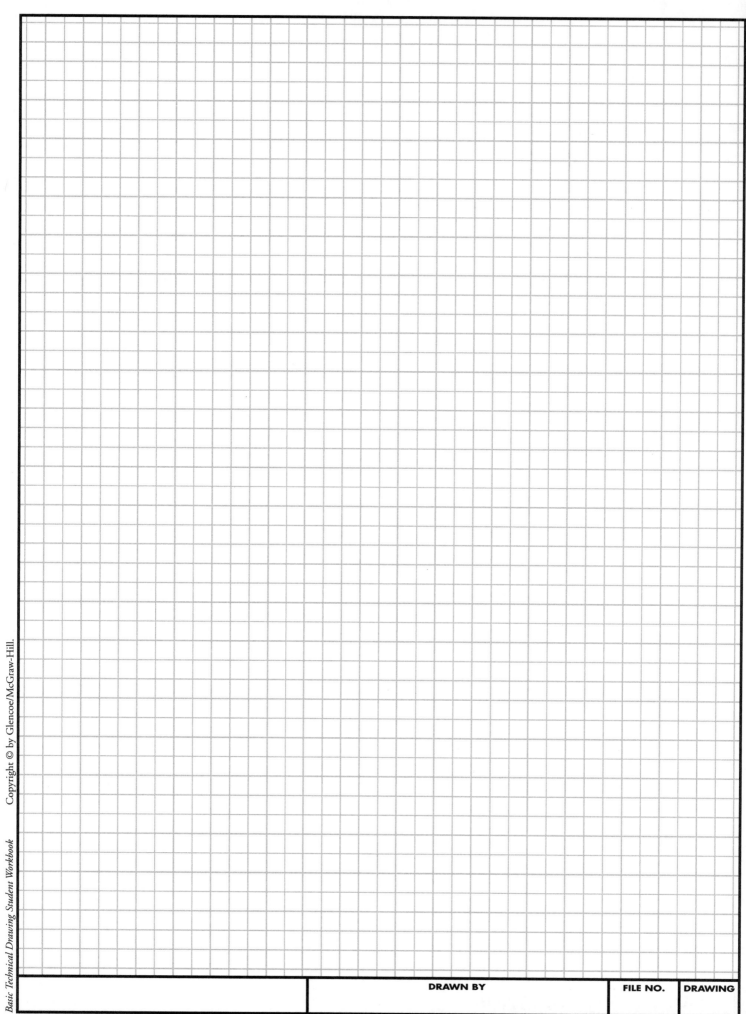

Basic Technical Drawing Student Workbook

		DRAWN BY	FILE NO.	DRAWING

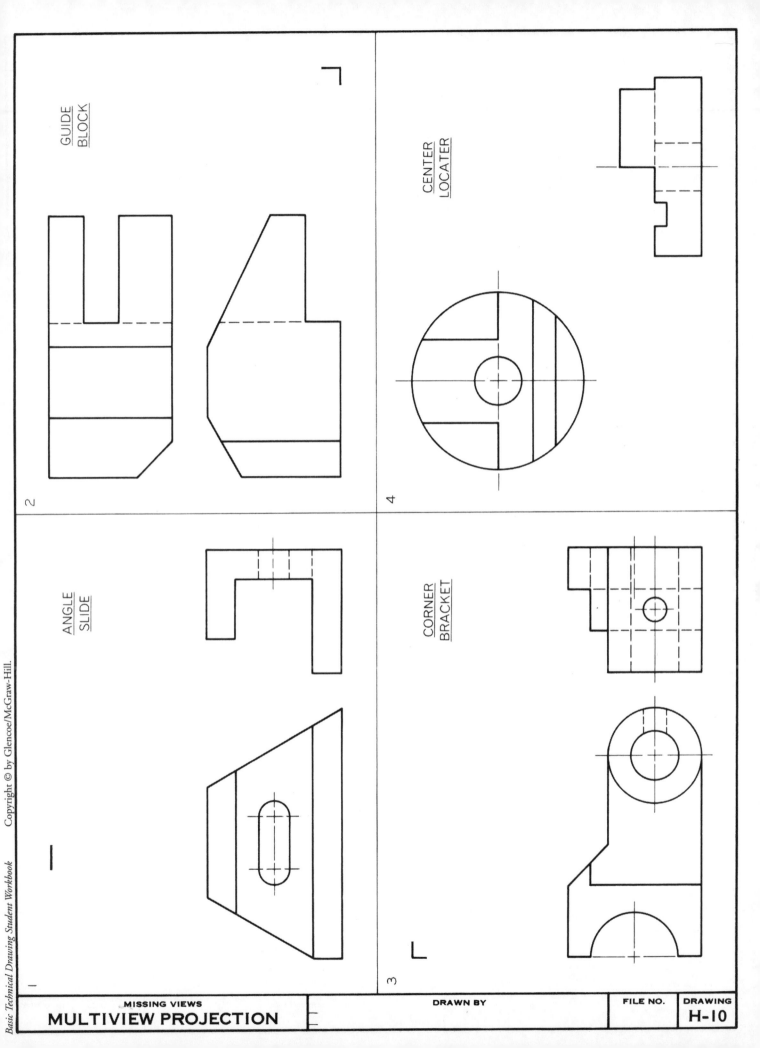

GUIDE
BLOCK

2

CENTER
LOCATER

4

ANGLE
SLIDE

1

CORNER
BRACKET

3

MISSING VIEWS

MULTIVIEW PROJECTION

DRAWN BY

FILE NO.

DRAWING

H-10

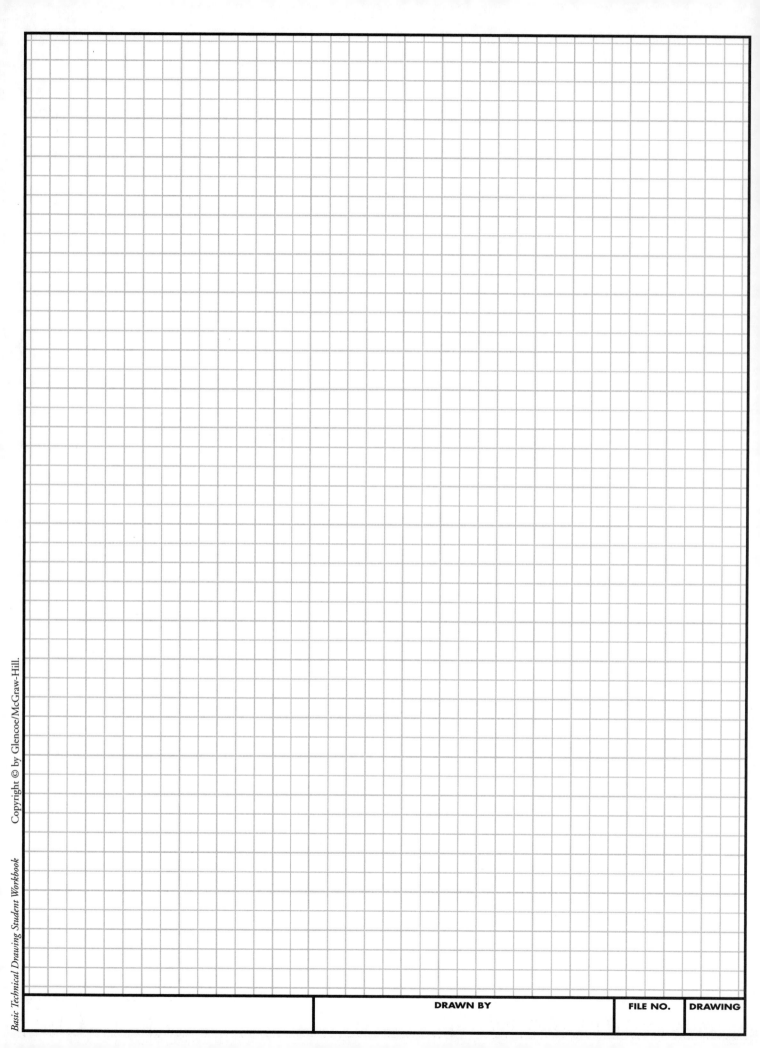

DRAWN BY

FILE NO.

DRAWING

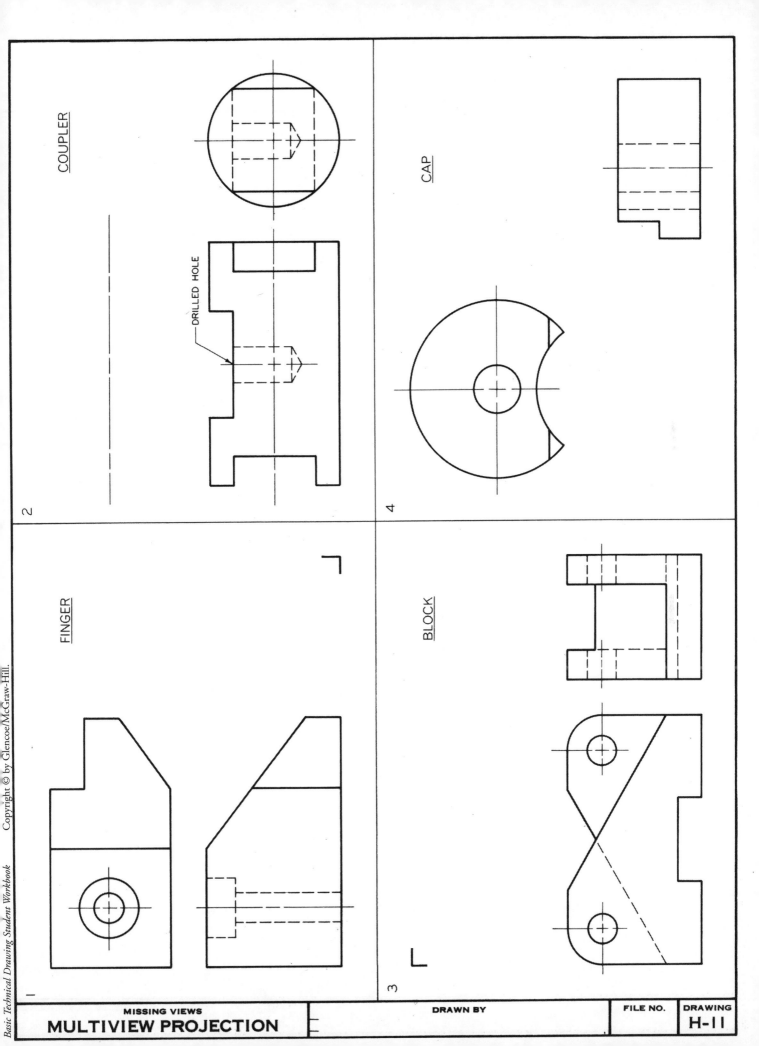

COUPLER

DRILLED HOLE

2

CAP

4

FINGER

BLOCK

3

MISSING VIEWS

MULTIVIEW PROJECTION

DRAWN BY

FILE NO.

DRAWING

H-11

		DRAWN BY		FILE NO.	DRAWING

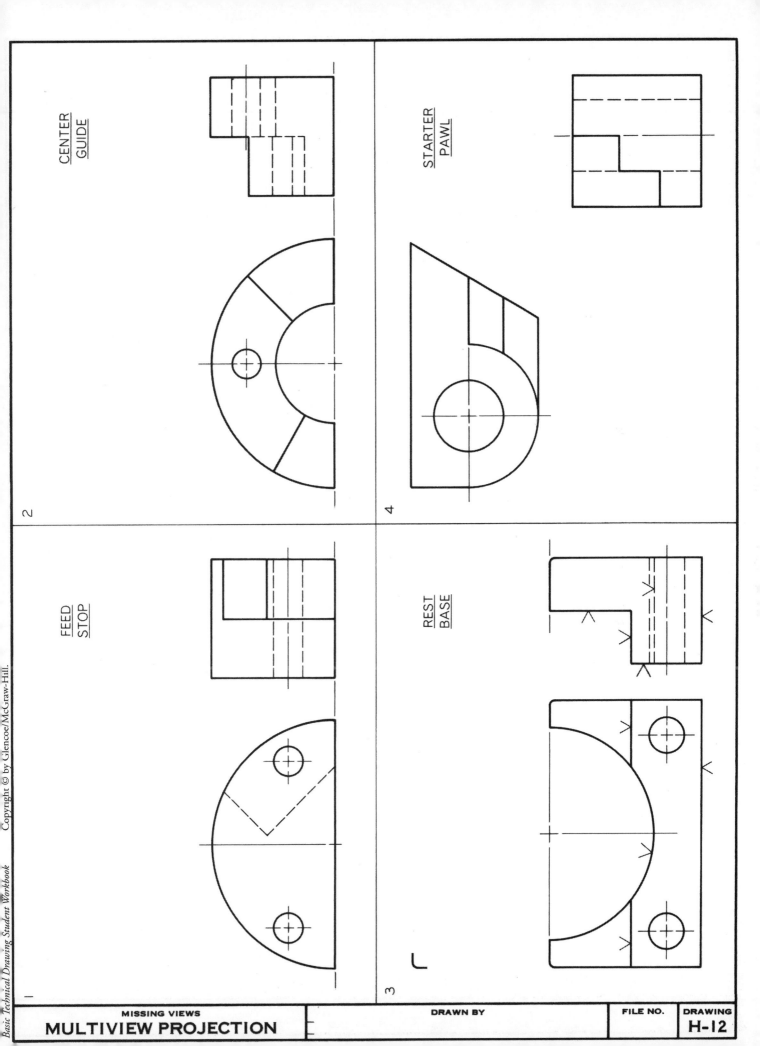

CENTER
GUIDE

STARTER
PAWL

FEED
STOP

REST
BASE

1

2

3

4

DRAWN BY

FILE NO.

DRAWING
H-12

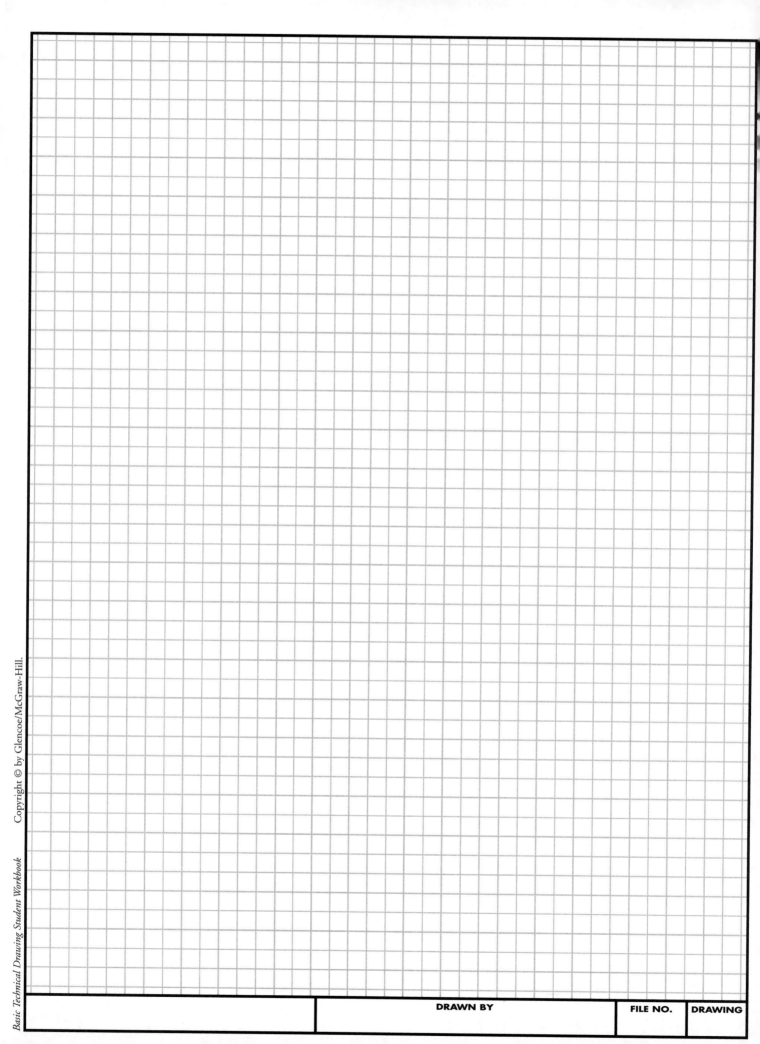

| | | DRAWN BY | | FILE NO. | DRAWING |

Draw front, top and
right side views.

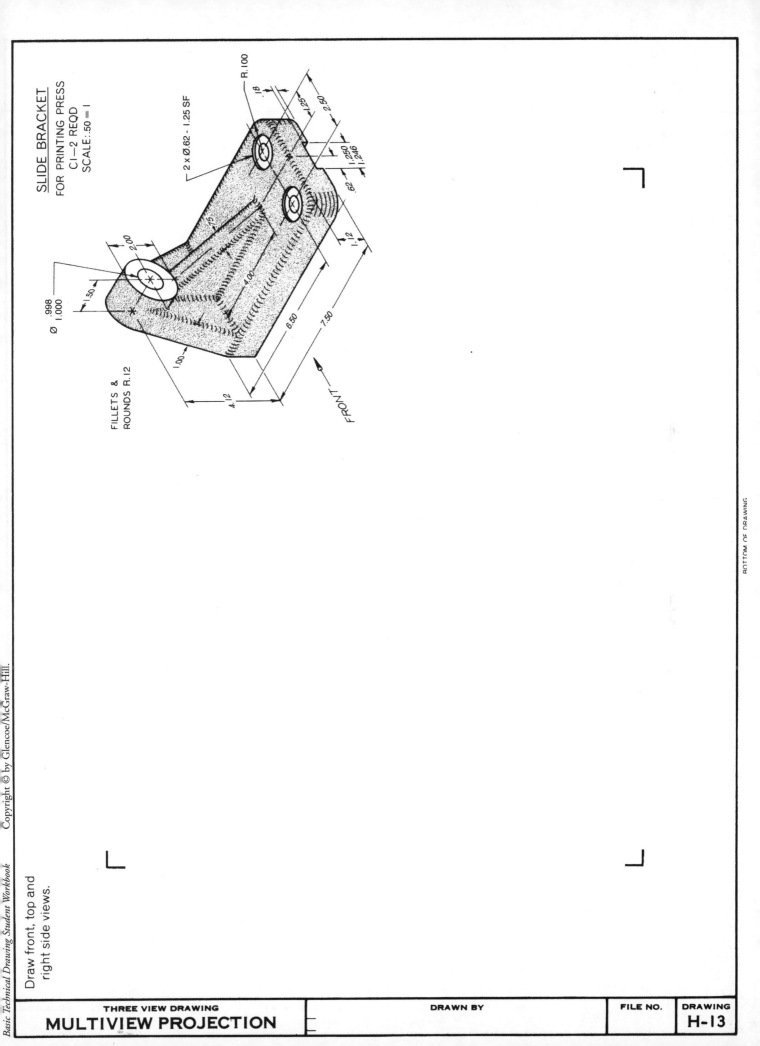

SLIDE BRACKET
FOR PRINTING PRESS
CI-2 REQD
SCALE:.50 = I

2 x Ø.62 - 1.25 SF

R.100

.18

1.25

2.50

1.250
1.246

.62

.12

1.12

4.00

6.50

7.50

.75

2.00

1.50

.998
Ø 1.000

1.00

4.12

FILLETS &
ROUNDS R.12

FRONT

BOTTOM OF DRAWING

THREE VIEW DRAWING	DRAWN BY	FILE NO.	DRAWING
MULTIVIEW PROJECTION			**H-13**

| | DRAWN BY | FILE NO. | DRAWING |

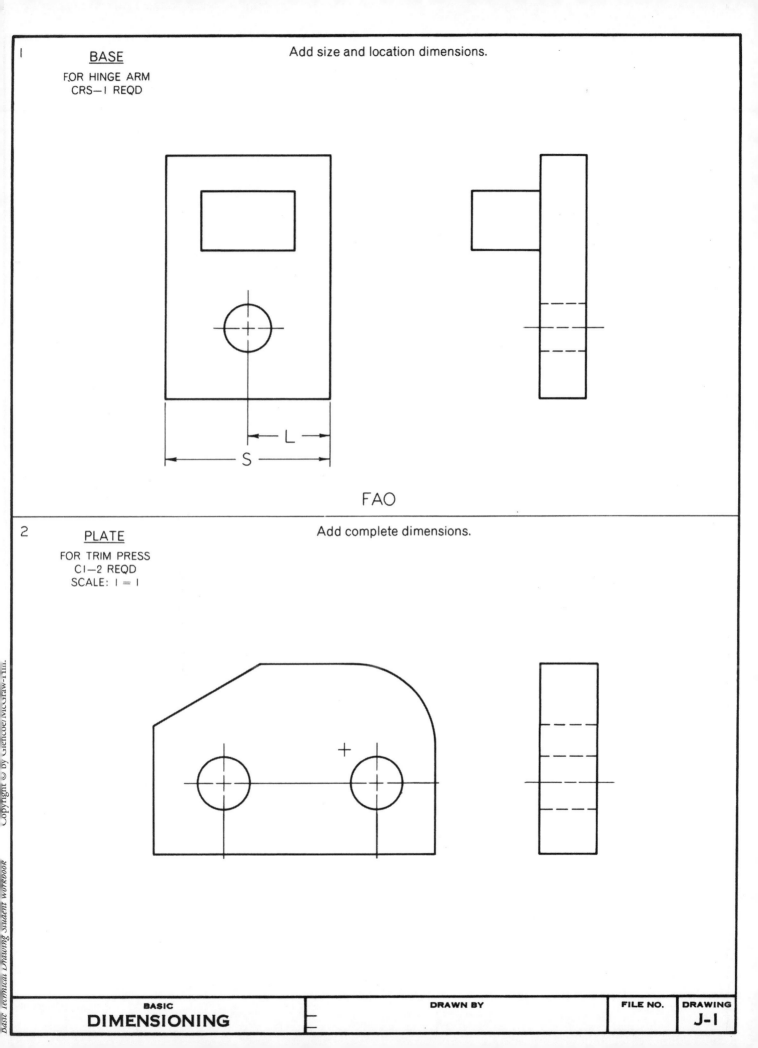

Add size and location dimensions.

1 BASE
FOR HINGE ARM
CRS—1 REQD

L

S

FAO

2 PLATE
FOR TRIM PRESS
CI—2 REQD
SCALE: 1 = 1

Add complete dimensions.

	DRAWN BY	FILE NO.	DRAWING

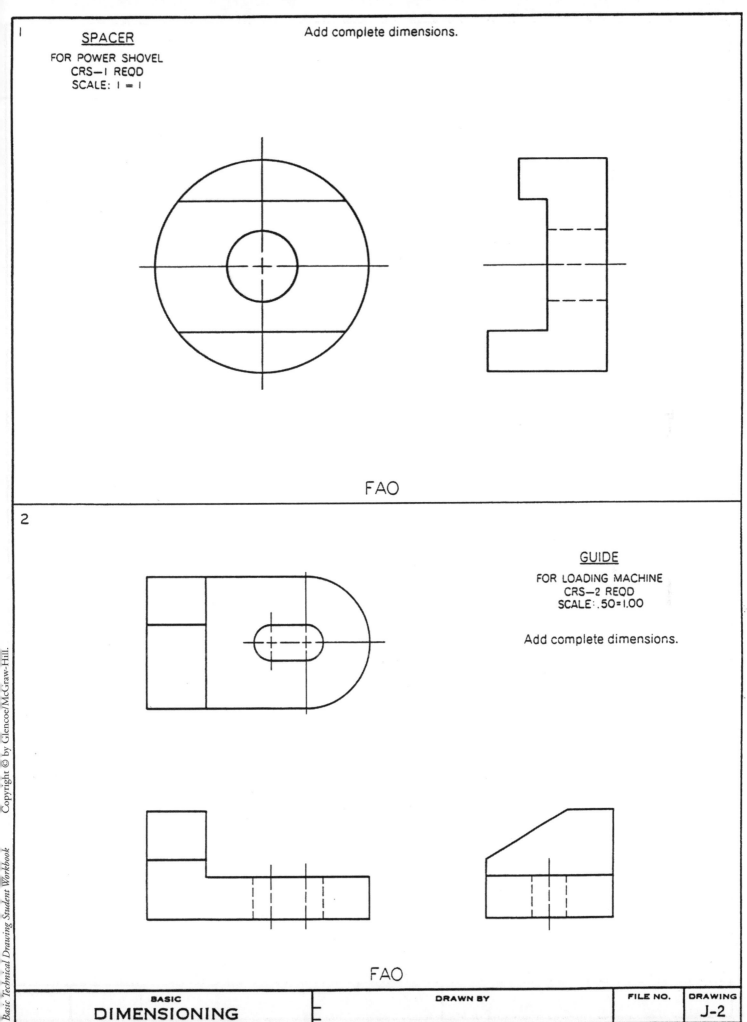

1 <u>SPACER</u>

FOR POWER SHOVEL
CRS—I REQD
SCALE: I = I

Add complete dimensions.

FAO

2

<u>GUIDE</u>

FOR LOADING MACHINE
CRS—2 REQD
SCALE: .50=1.00

Add complete dimensions.

FAO

BASIC DIMENSIONING		DRAWN BY	FILE NO.	DRAWING J-2

Basic Technical Drawing Student Workbook

		DRAWN BY		FILE NO.	DRAWING

Add complete dimensions.

1

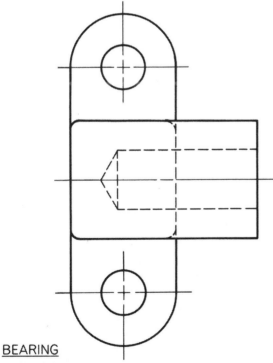

<u>SHIFTER</u>

FOR RIVET MACHINE
CI—I REQD
SCALE: I = I

2

Add complete dimensions.

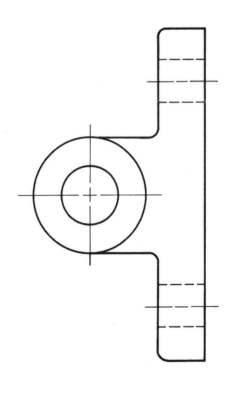

<u>BEARING</u>

FOR AUTOMATIC HOIST
CI—2 REQD
SCALE: ½ = I

BASIC **DIMENSIONING**		DRAWN BY	FILE NO.	DRAWING **J-3**

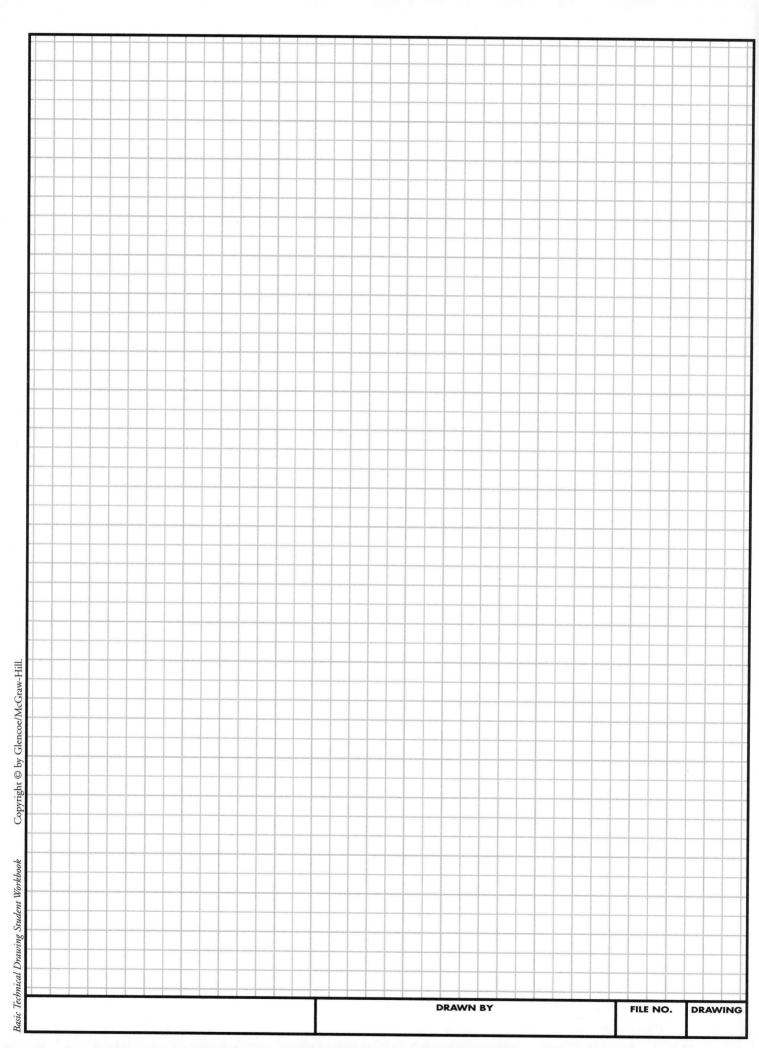

Basic Technical Drawing Student Workbook

		DRAWN BY		FILE NO.	DRAWING

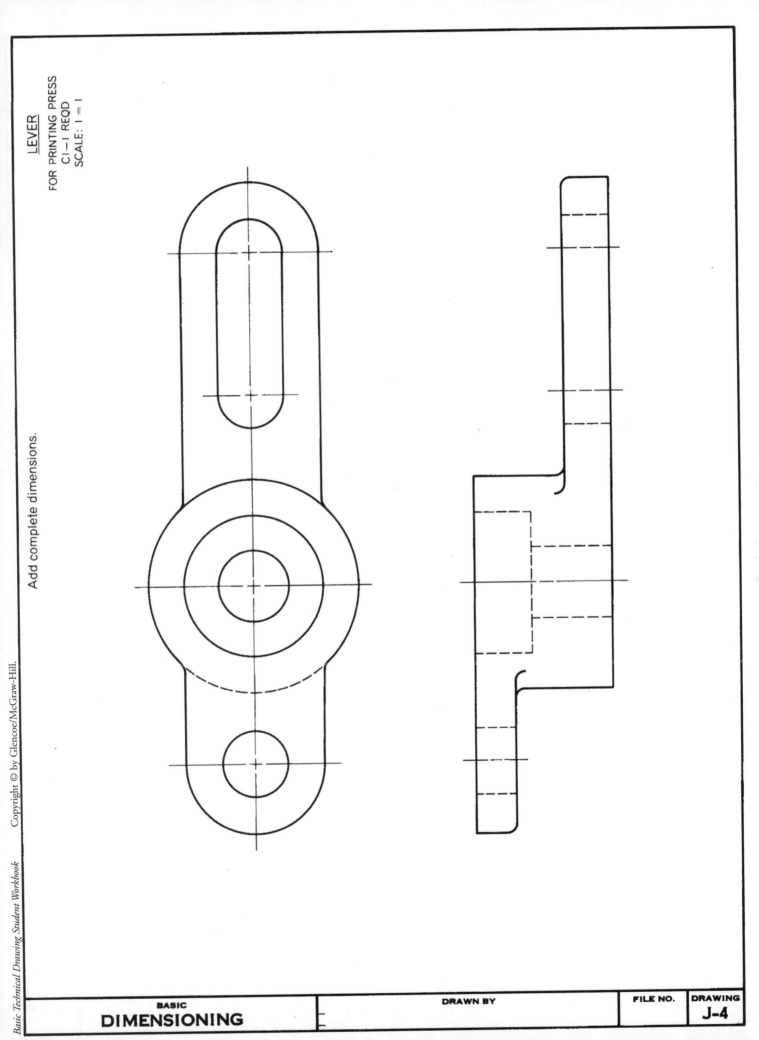

Add complete dimensions.

<u>LEVER</u>

FOR PRINTING PRESS
C1 – 1 REQD
SCALE: 1 = 1

BASIC
DIMENSIONING

DRAWN BY

FILE NO.

DRAWING
J-4

| | DRAWN BY | FILE NO. | DRAWING |

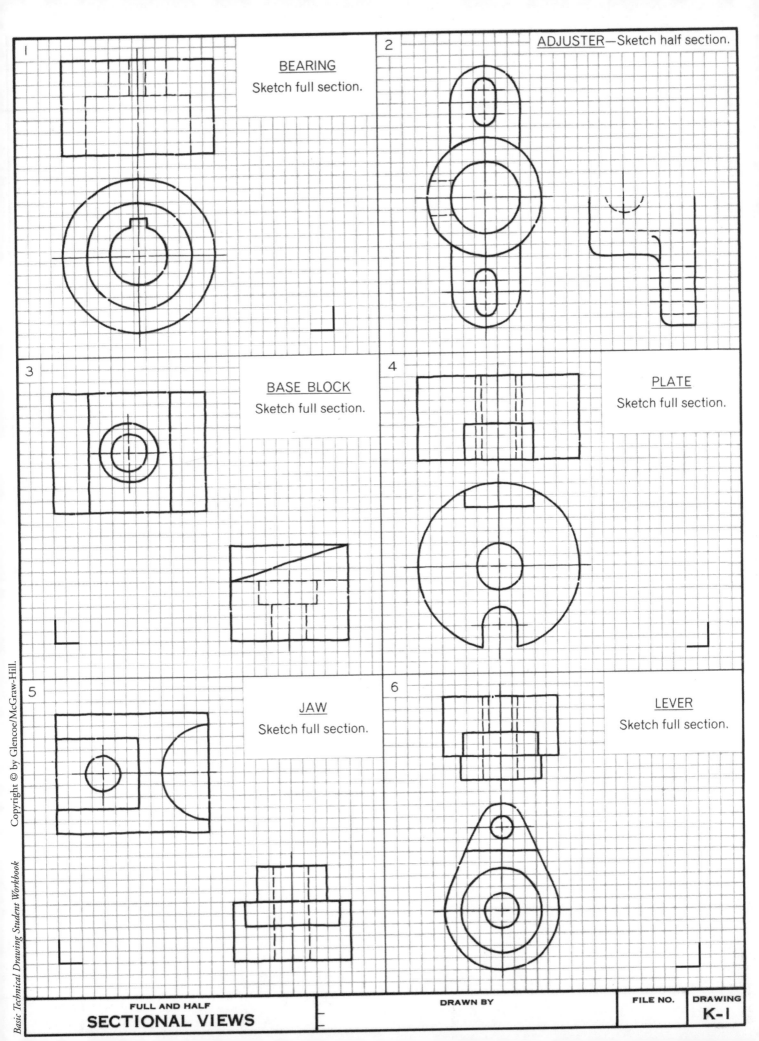

1 **BEARING**
Sketch full section.

2 ADJUSTER—Sketch half section.

3 **BASE BLOCK**
Sketch full section.

4 **PLATE**
Sketch full section.

5 **JAW**
Sketch full section.

6 **LEVER**
Sketch full section.

FULL AND HALF
SECTIONAL VIEWS

DRAWN BY

FILE NO.

DRAWING
K-1

		DRAWN BY	FILE NO.	DRAWING

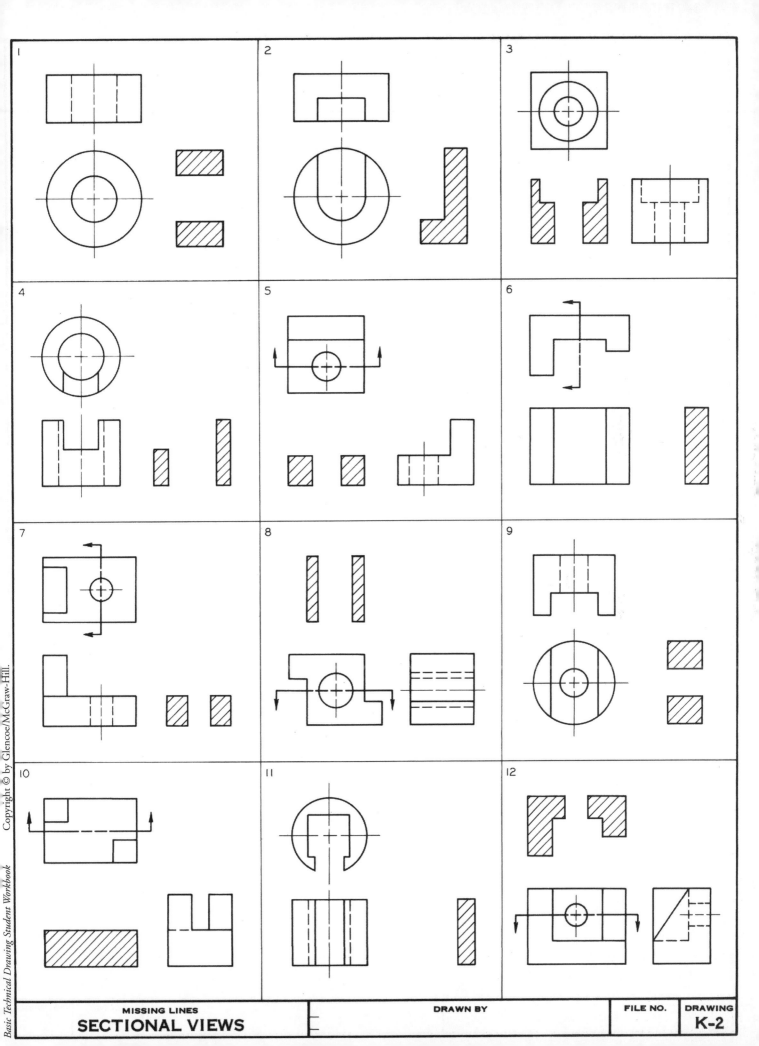

| MISSING LINES | DRAWN BY | FILE NO. | DRAWING |
| SECTIONAL VIEWS | | | K-2 |

		DRAWN BY		FILE NO.	DRAWING

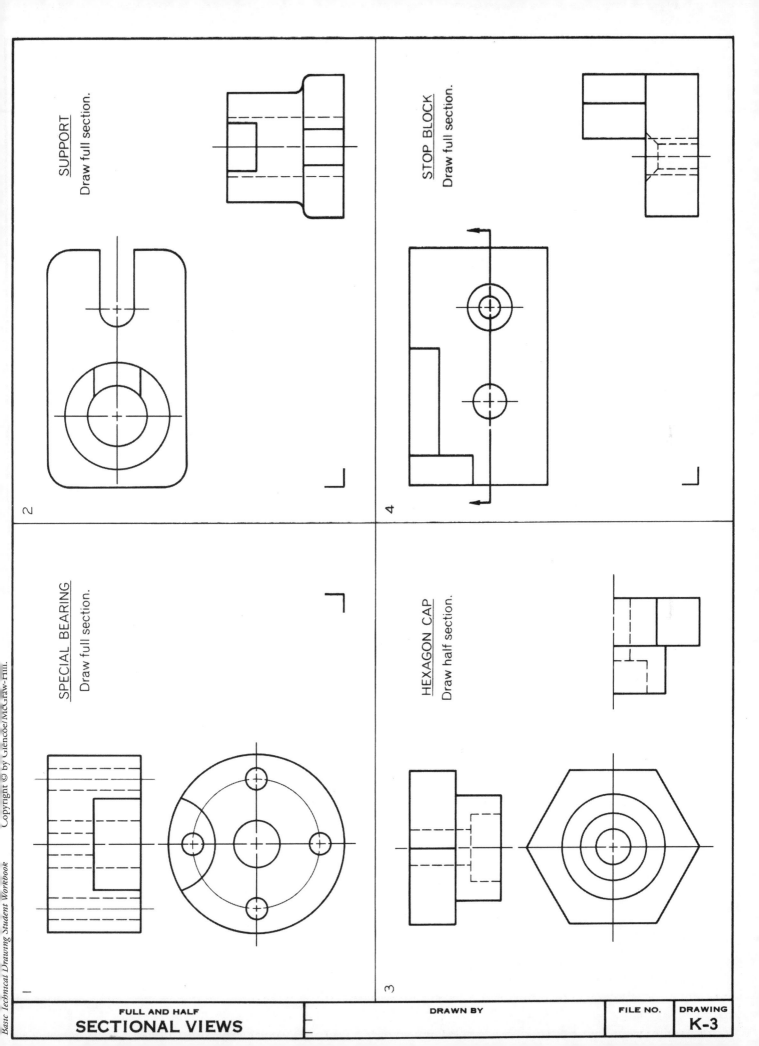

SUPPORT
Draw full section.

STOP BLOCK
Draw full section.

SPECIAL BEARING
Draw full section.

HEXAGON CAP
Draw half section.

2

4

1

3

FULL AND HALF
SECTIONAL VIEWS

DRAWN BY

FILE NO.

DRAWING
K-3

| | | | **DRAWN BY** | | **FILE NO.** | **DRAWING** |

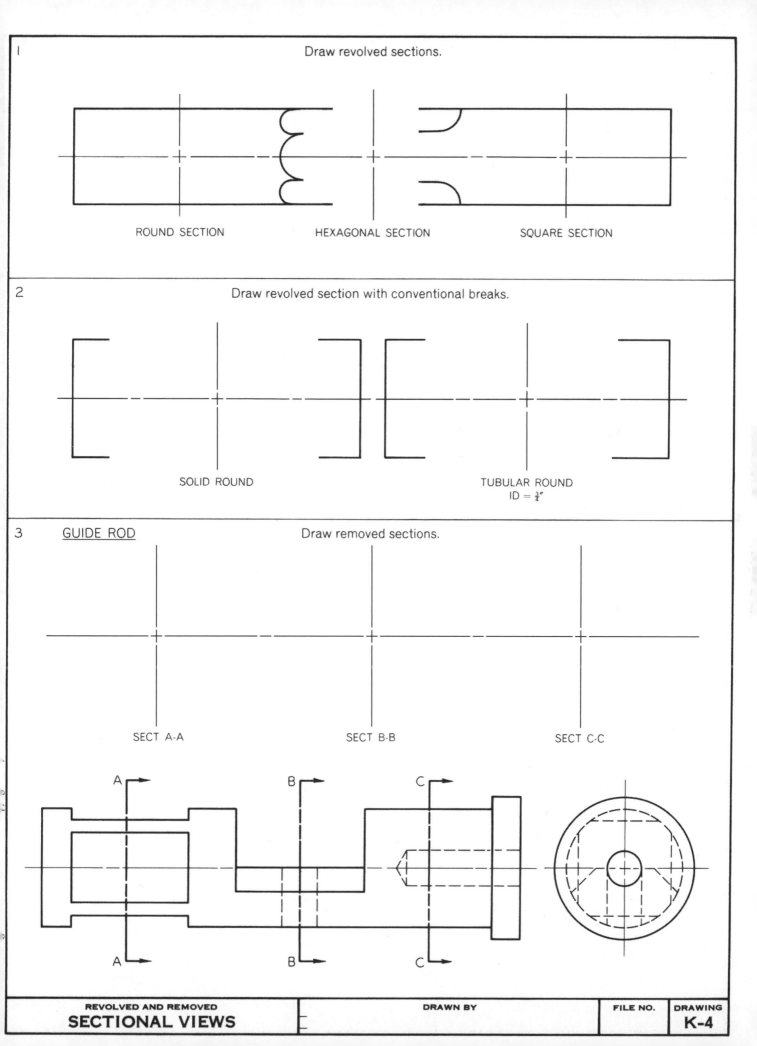

1   Draw revolved sections.

ROUND SECTION   HEXAGONAL SECTION   SQUARE SECTION

2   Draw revolved section with conventional breaks.

SOLID ROUND   TUBULAR ROUND
ID = $\frac{3}{4}''$

3   GUIDE ROD   Draw removed sections.

SECT A-A   SECT B-B   SECT C-C

A   B   C

A   B   C

REVOLVED AND REMOVED	DRAWN BY	FILE NO.	DRAWING
SECTIONAL VIEWS			**K-4**

Basic Technical Drawing Student Workbook

		DRAWN BY		FILE NO.	DRAWING

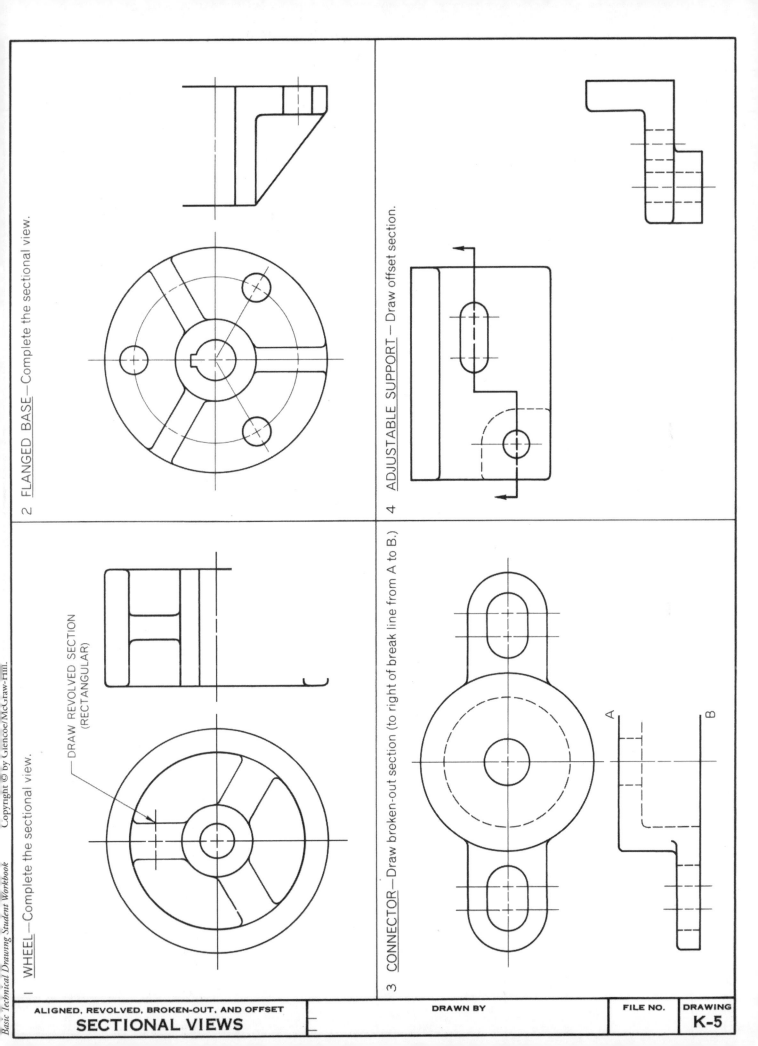

2 FLANGED BASE—Complete the sectional view.

4 ADJUSTABLE SUPPORT—Draw offset section.

1 WHEEL—Complete the sectional view.

DRAW REVOLVED SECTION
(RECTANGULAR)

3 CONNECTOR—Draw broken-out section (to right of break line from A to B.)

A

B

ALIGNED, REVOLVED, BROKEN-OUT, AND OFFSET
SECTIONAL VIEWS

DRAWN BY

FILE NO.

DRAWING
K-5

		DRAWN BY		FILE NO.	DRAWING

Basic Technical Drawing Student Workbook

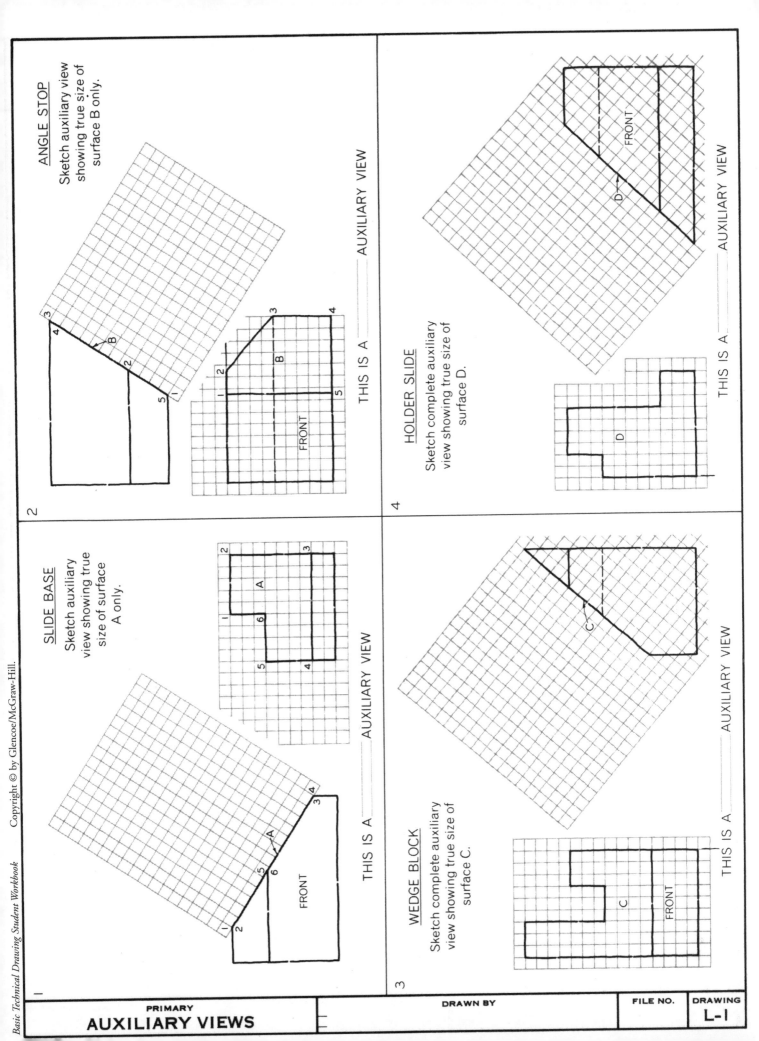

2

ANGLE STOP

Sketch auxiliary view showing true size of surface B only.

THIS IS A ———— AUXILIARY VIEW

1

SLIDE BASE

Sketch auxiliary view showing true size of surface A only.

THIS IS A ———— AUXILIARY VIEW

4

HOLDER SLIDE

Sketch complete auxiliary view showing true size of surface D.

THIS IS A ———— AUXILIARY VIEW

3

WEDGE BLOCK

Sketch complete auxiliary view showing true size of surface C.

THIS IS A ———— AUXILIARY VIEW

PRIMARY
AUXILIARY VIEWS

DRAWN BY

FILE NO.

DRAWING
L-1

		DRAWN BY	FILE NO.	DRAWING

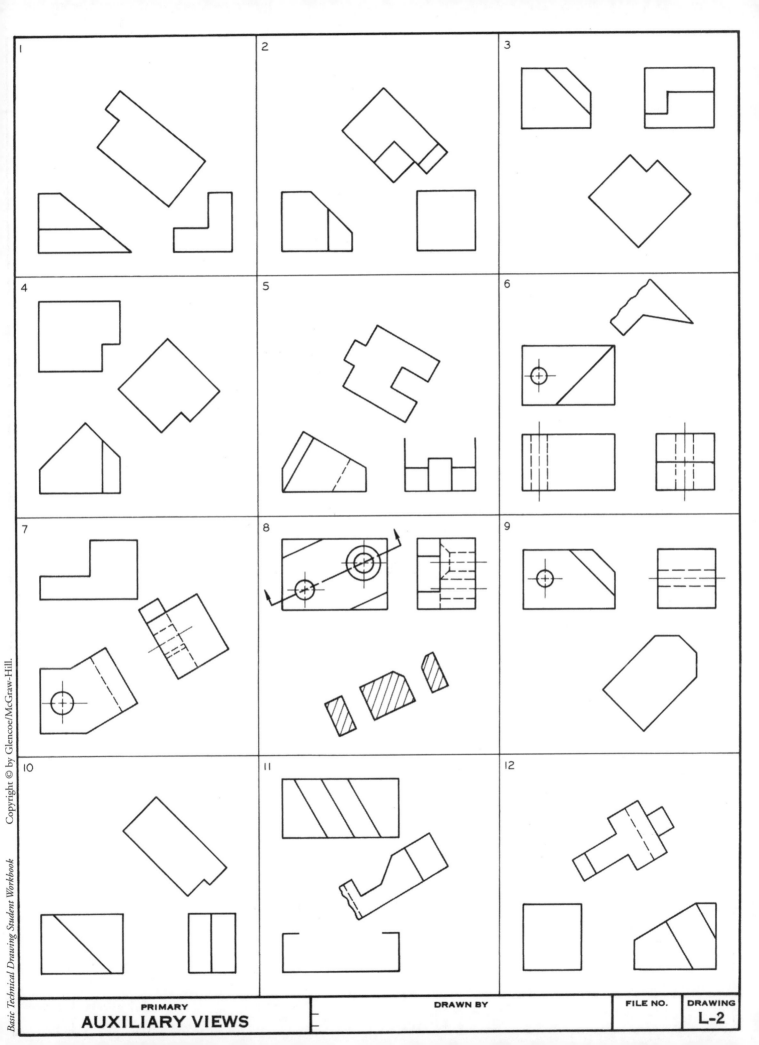

PRIMARY
AUXILIARY VIEWS

DRAWN BY

FILE NO.

DRAWING
L-2

	DRAWN BY	FILE NO.	DRAWING

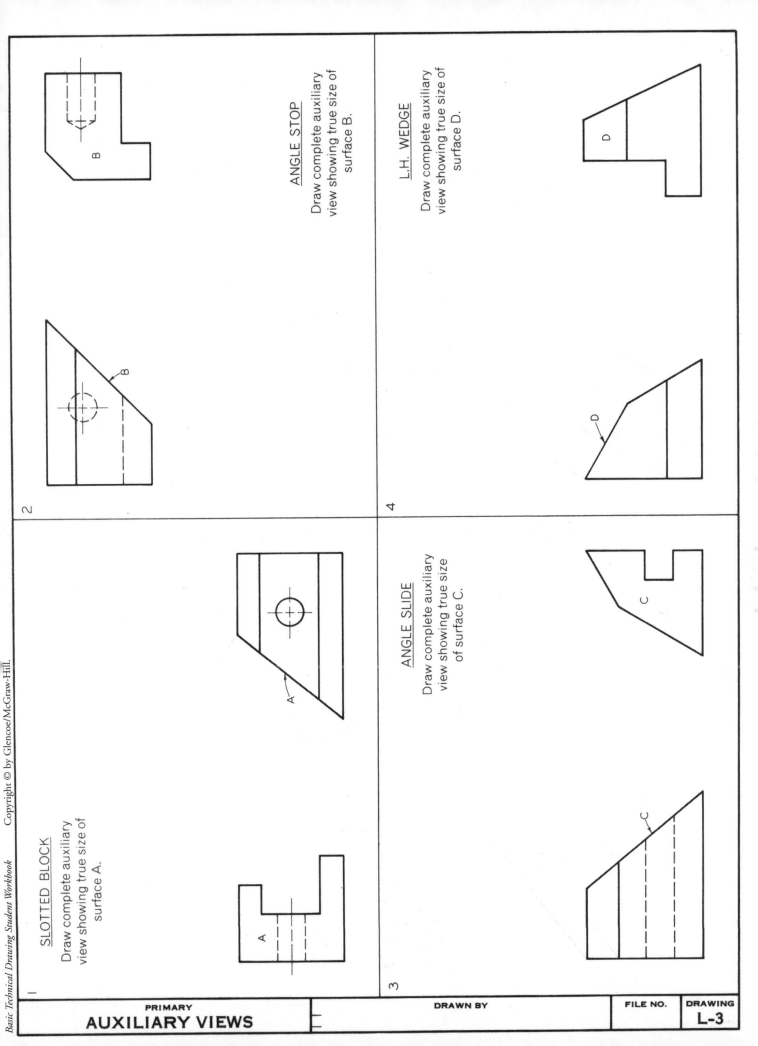

1

SLOTTED BLOCK

Draw complete auxiliary
view showing true size of
surface A.

A

2

3

ANGLE SLIDE

Draw complete auxiliary
view showing true size
of surface C.

C

4

ANGLE STOP

Draw complete auxiliary
view showing true size of
surface B.

B

L.H. WEDGE

Draw complete auxiliary
view showing true size of
surface D.

D

PRIMARY
AUXILIARY VIEWS

DRAWN BY

FILE NO.

DRAWING
L-3

		DRAWN BY		FILE NO.	DRAWING

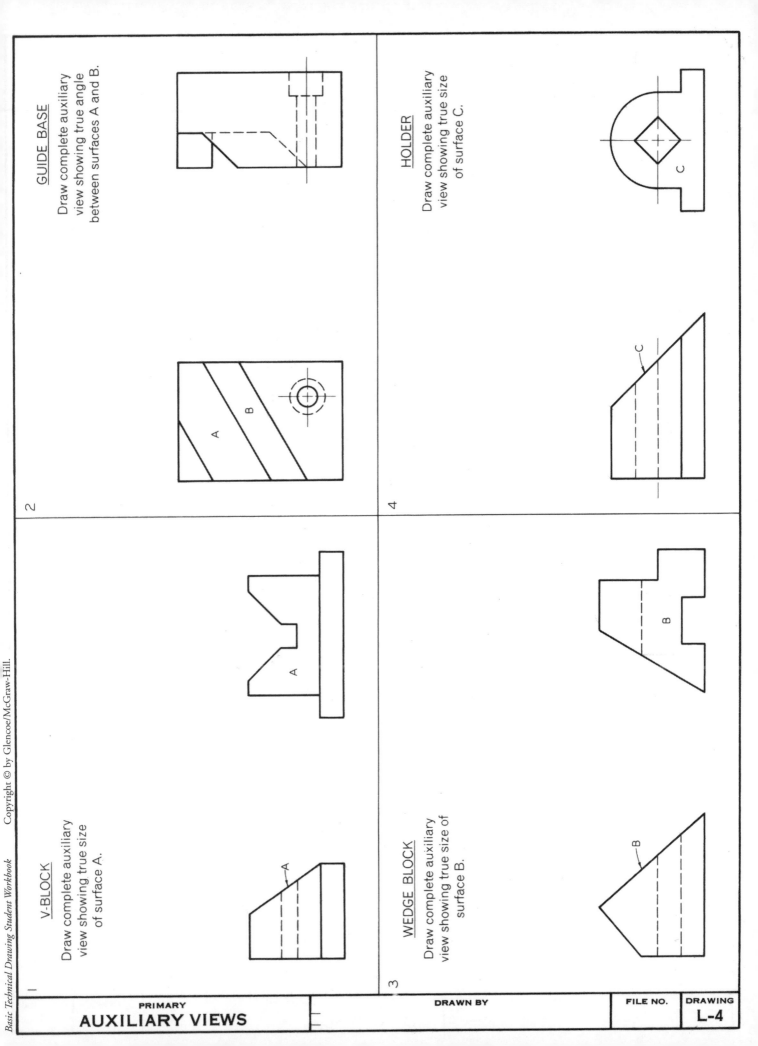

GUIDE BASE

Draw complete auxiliary view showing true angle between surfaces A and B.

2

HOLDER

Draw complete auxiliary view showing true size of surface C.

C

4

V-BLOCK

Draw complete auxiliary view showing true size of surface A.

A

A

3

WEDGE BLOCK

Draw complete auxiliary view showing true size of surface B.

B

B

1

PRIMARY
AUXILIARY VIEWS

DRAWN BY

FILE NO.   DRAWING
L-4

Basic Technical Drawing Student Workbook

DRAWN BY

FILE NO.

DRAWING

1

Γ

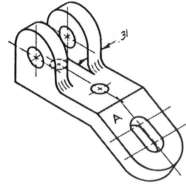

.31

ANGLE ARM

Draw partial top view, and
partial auxiliary view
showing true size of
surface A.

A

2

Ø.31-⊔Ø.62 ↧.19

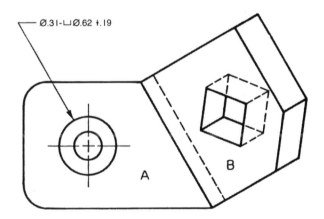

A   B

BRACKET

Draw complete primary auxiliary
view showing 140° angle between
surfaces A and B; then draw partial
secondary auxiliary view showing
true size of member B. Thickness
of members = .44.

		DRAWN BY		FILE NO.	DRAWING

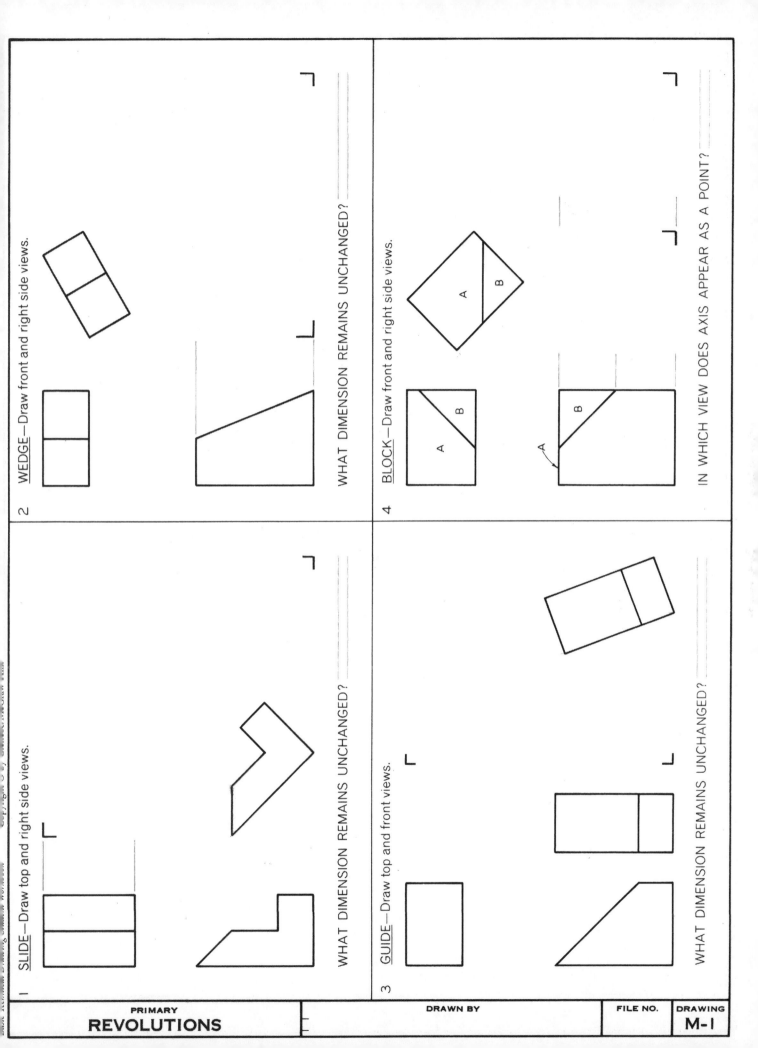

1 SLIDE—Draw top and right side views.

WHAT DIMENSION REMAINS UNCHANGED? _____

2 WEDGE—Draw front and right side views.

WHAT DIMENSION REMAINS UNCHANGED? _____

3 GUIDE—Draw top and front views.

WHAT DIMENSION REMAINS UNCHANGED? _____

4 BLOCK—Draw front and right side views.

A B

A B

A B

IN WHICH VIEW DOES AXIS APPEAR AS A POINT? _____

PRIMARY
REVOLUTIONS

DRAWN BY

FILE NO.

DRAWING
M-I

		DRAWN BY	FILE NO.	DRAWING

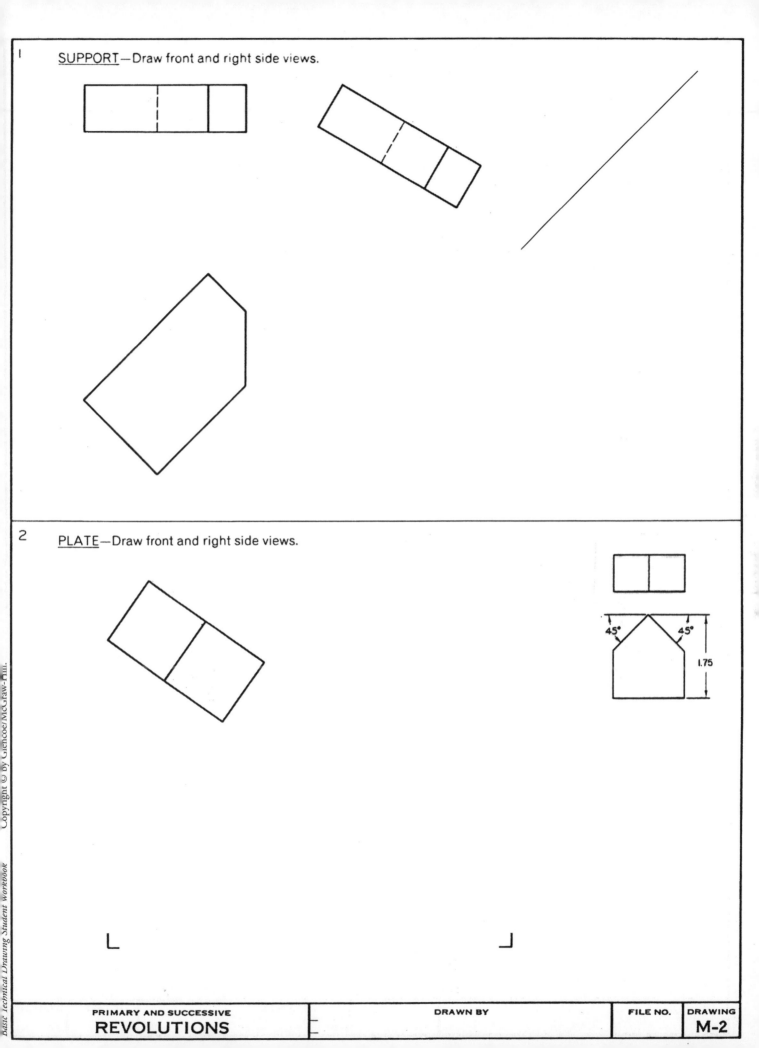

1 <u>SUPPORT</u>—Draw front and right side views.

2 <u>PLATE</u>—Draw front and right side views.

45°   45°

1.75

PRIMARY AND SUCCESSIVE
REVOLUTIONS

DRAWN BY

FILE NO.

DRAWING
M-2

| | | DRAWN BY | | FILE NO. | DRAWING |

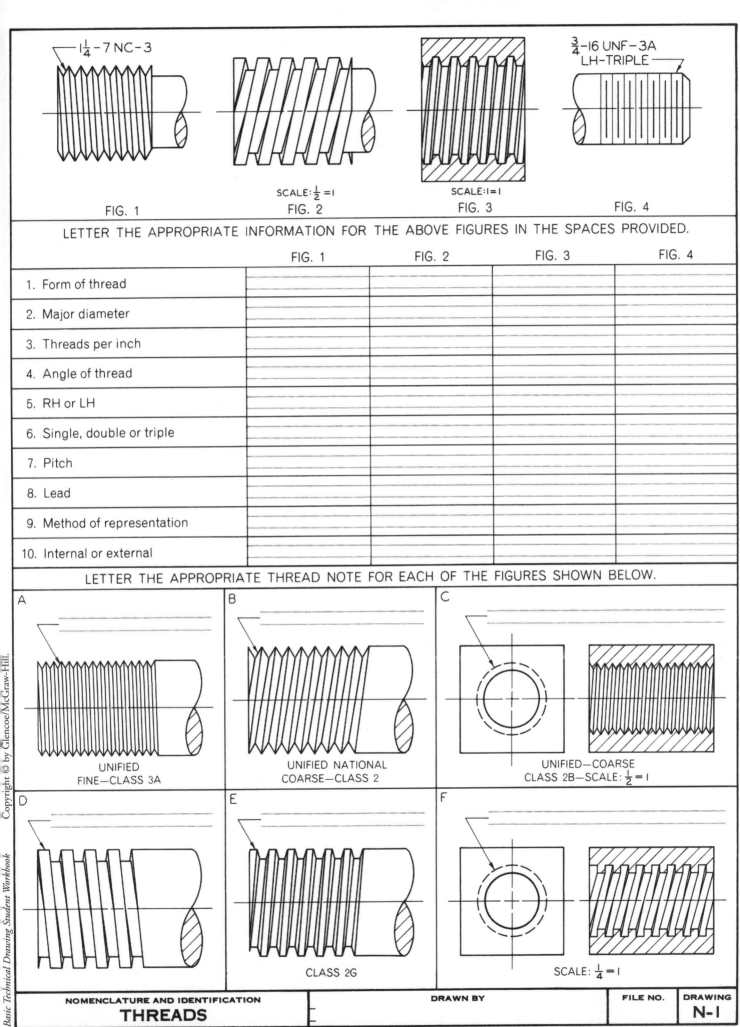

LETTER THE APPROPRIATE INFORMATION FOR THE ABOVE FIGURES IN THE SPACES PROVIDED.

	FIG. 1	FIG. 2	FIG. 3	FIG. 4
1. Form of thread				
2. Major diameter				
3. Threads per inch				
4. Angle of thread				
5. RH or LH				
6. Single, double or triple				
7. Pitch				
8. Lead				
9. Method of representation				
10. Internal or external				

LETTER THE APPROPRIATE THREAD NOTE FOR EACH OF THE FIGURES SHOWN BELOW.

A — UNIFIED FINE—CLASS 3A

B — UNIFIED NATIONAL COARSE—CLASS 2

C — UNIFIED—COARSE CLASS 2B—SCALE: $\frac{1}{2}$ = 1

D

E — CLASS 2G

F — SCALE: $\frac{1}{4}$ = 1

NOMENCLATURE AND IDENTIFICATION
THREADS

DRAWN BY

FILE NO.

DRAWING N-1

Basic Technical Drawing Student Workbook

DRAWN BY

FILE NO.

DRAWING

Complete the view.

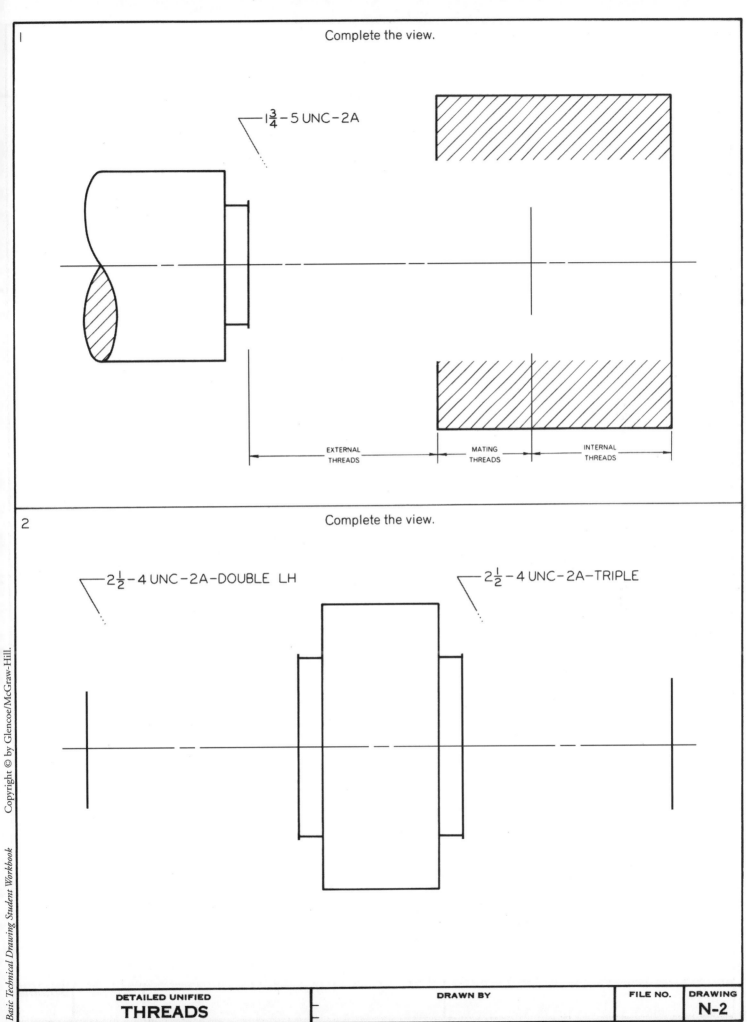

1

$1\frac{3}{4} - 5$ UNC – 2A

EXTERNAL
THREADS

MATING
THREADS

INTERNAL
THREADS

Complete the view.

2

$2\frac{1}{2} - 4$ UNC – 2A – DOUBLE LH

$2\frac{1}{2} - 4$ UNC – 2A – TRIPLE

DETAILED UNIFIED **THREADS**		DRAWN BY	FILE NO.	DRAWING **N-2**

		DRAWN BY		FILE NO.	DRAWING

Complete the view.

1

$4\frac{1}{2}-1$ SQUARE

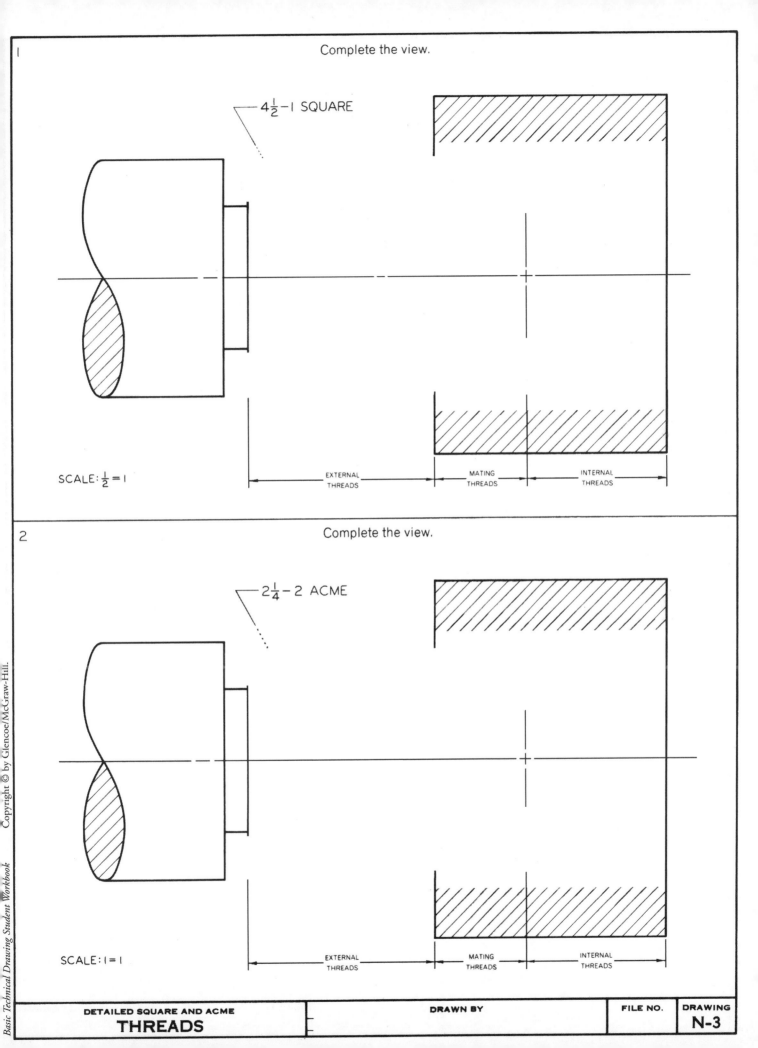

SCALE: $\frac{1}{2}=1$

EXTERNAL THREADS   MATING THREADS   INTERNAL THREADS

Complete the view.

2

$2\frac{1}{4}-2$ ACME

SCALE: $1=1$

EXTERNAL THREADS   MATING THREADS   INTERNAL THREADS

DETAILED SQUARE AND ACME
THREADS

DRAWN BY

FILE NO.

DRAWING
N-3

Basic Technical Drawing Student Workbook

		DRAWN BY	FILE NO.	DRAWING

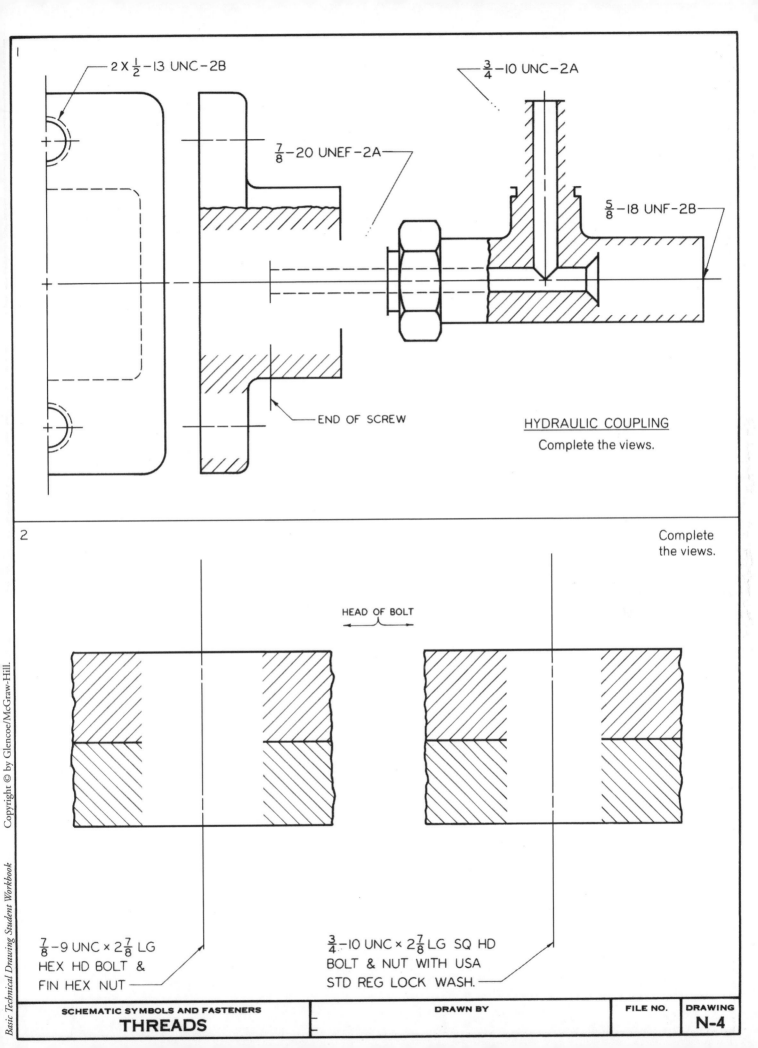

1

$2 \times \frac{1}{2} - 13$ UNC − 2B

$\frac{3}{4} - 10$ UNC − 2A

$\frac{7}{8} - 20$ UNEF − 2A

$\frac{5}{8} - 18$ UNF − 2B

END OF SCREW

HYDRAULIC COUPLING

Complete the views.

2

Complete the views.

HEAD OF BOLT

$\frac{7}{8} - 9$ UNC × $2\frac{7}{8}$ LG
HEX HD BOLT &
FIN HEX NUT

$\frac{3}{4} - 10$ UNC × $2\frac{7}{8}$ LG SQ HD
BOLT & NUT WITH USA
STD REG LOCK WASH.

SCHEMATIC SYMBOLS AND FASTENERS	DRAWN BY	FILE NO.	DRAWING
THREADS			**N-4**

Basic Technical Drawing Student Workbook

	DRAWN BY	FILE NO.	DRAWING

1. Complete the sectioned assembly as specified.

① $\frac{5}{16} \times 1"$ SLOT. FLAT HD CAP SCR

② $\frac{3}{8} \times 1"$ SLOT. FLAT PT SET SCR

③ $\frac{3}{8} \times 1\frac{1}{16}$ HEX HD CAP SCR & USA STD REG LOCKWASHER

④ $\frac{5}{16} \times \frac{5}{8}$ RD HD MACH SCR

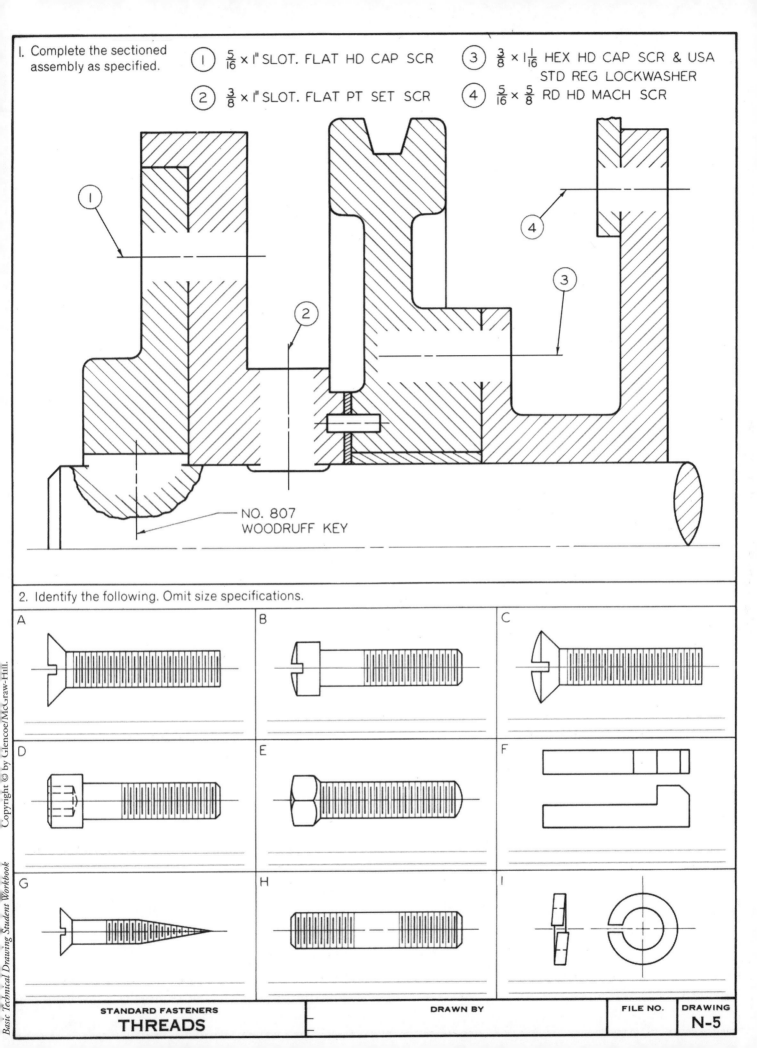

NO. 807
WOODRUFF KEY

2. Identify the following. Omit size specifications.

A

B

C

D

E

F

G

H

I

| | | DRAWN BY | | FILE NO. | DRAWING |

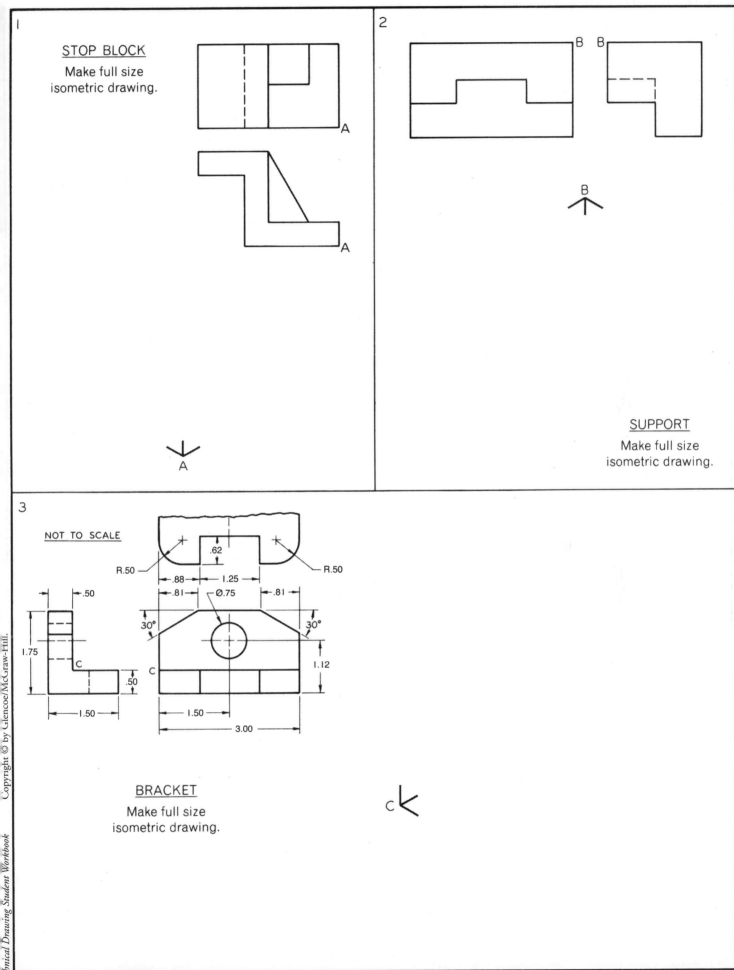

1

STOP BLOCK
Make full size
isometric drawing.

A

A

A

2

B B

B B

B

SUPPORT
Make full size
isometric drawing.

3

NOT TO SCALE

R.50

R.50

.62

.88

1.25

.81

Ø.75

.81

.50

30°

30°

1.12

1.75

C

C

.50

1.50

1.50

3.00

BRACKET
Make full size
isometric drawing.

C

		DRAWN BY		FILE NO.	DRAWING

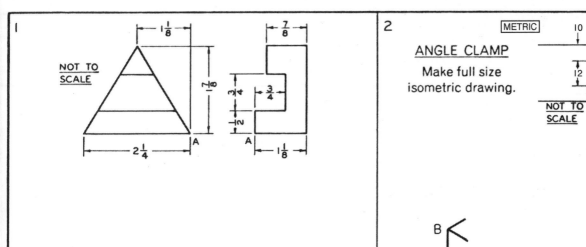

1

NOT TO SCALE

$1\frac{1}{8}$

$2\frac{1}{4}$

$\frac{7}{8}$

$1\frac{7}{8}$

$\frac{3}{4}$

$\frac{3}{4}$

$\frac{1}{2}$

$1\frac{1}{8}$

A

A

SLIDE KEY

Make full size
isometric drawing.

A

2

METRIC

ANGLE CLAMP

Make full size
isometric drawing.

NOT TO SCALE

10

12

16

30°

12

30°

B

56

28

B

40°

32

B

3

2.50

1.25

1.25

2.50

.38

C

2.00

.76

.38

.62

2.25

.50

C

NOT TO SCALE

SAFETY KEY

Make full size
isometric drawing.

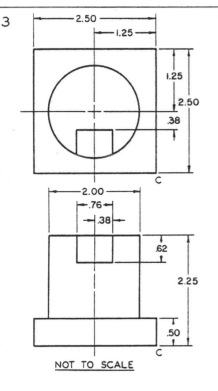

C

PICTORIAL DRAWING		DRAWN BY	FILE NO.	DRAWING
ISOMETRIC DRAWING				**O-2**

		DRAWN BY		FILE NO.	DRAWING

1

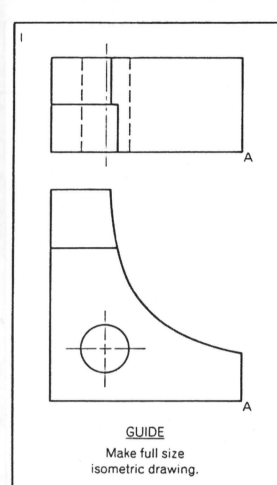

<u>GUIDE</u>

Make full size
isometric drawing.

A

2

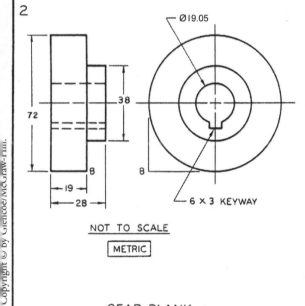

Ø19.05

38

72

8

8

19

28

6 × 3 KEYWAY

<u>NOT TO SCALE</u>

METRIC

<u>GEAR BLANK</u>

Make full size
isometric drawing.

B

PICTORIAL DRAWING		DRAWN BY	FILE NO.	DRAWING
ISOMETRIC DRAWING				**O-3**

| | | DRAWN BY | FILE NO. | DRAWING |

1

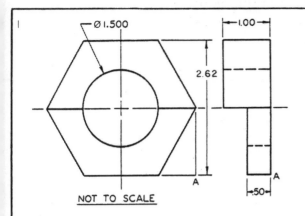

Ø1.500

1.00

2.62

A

NOT TO SCALE

.50

HEX BASE

Make full size
isometric drawing.

A

2

CENTER SLIDE

Make full size
isometric drawing.

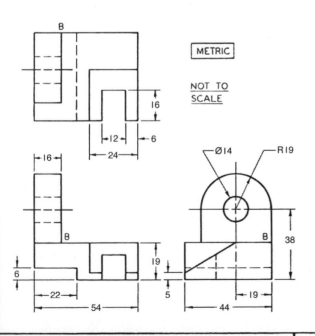

B

METRIC

NOT TO
SCALE

16

12   6

24

16

B

Ø14   R19

38

B

19

6

5   19

22

44

54

B

PICTORIAL DRAWING	DRAWN BY	FILE NO.	DRAWING
ISOMETRIC DRAWING			O-4

Basic Technical Drawing Student Workbook

	DRAWN BY	FILE NO.	DRAWING

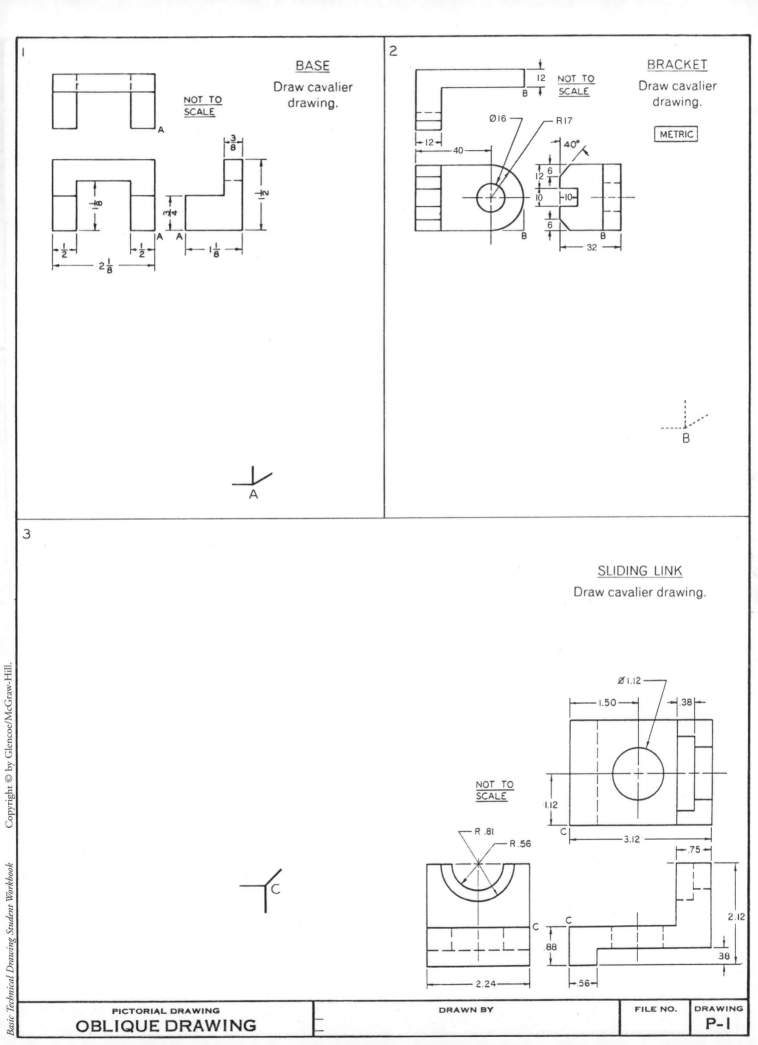

1

BASE
Draw cavalier drawing.

NOT TO SCALE

2

BRACKET
Draw cavalier drawing.

NOT TO SCALE

METRIC

3

SLIDING LINK
Draw cavalier drawing.

NOT TO SCALE

Basic Technical Drawing Student Workbook

DRAWN BY   **FILE NO.**   **DRAWING**

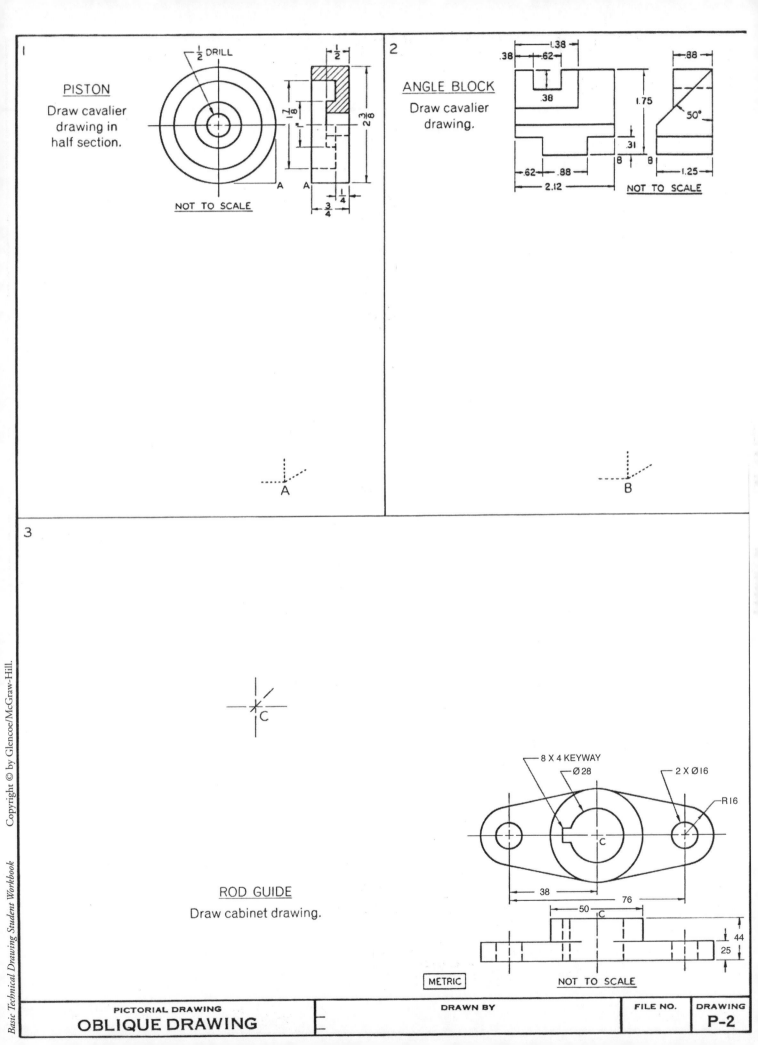

1

PISTON
Draw cavalier
drawing in
half section.

½ DRILL

NOT TO SCALE

A   A

½

2 3/8

1 7/8

3/4

1/4

A

2

ANGLE BLOCK
Draw cavalier
drawing.

1.38
.38   .62

.38

.62   .88

2.12

1.75

.31

8   8

.88

50°

1.25

NOT TO SCALE

B

3

C

ROD GUIDE

Draw cabinet drawing.

8 X 4 KEYWAY
Ø 28
2 X Ø16
R16

C

38

76

50   C

44

25

METRIC   NOT TO SCALE

PICTORIAL DRAWING		DRAWN BY	FILE NO.	DRAWING
OBLIQUE DRAWING				**P-2**

Basic Technical Drawing Student Workbook

DRAWN BY

FILE NO. **DRAWING**

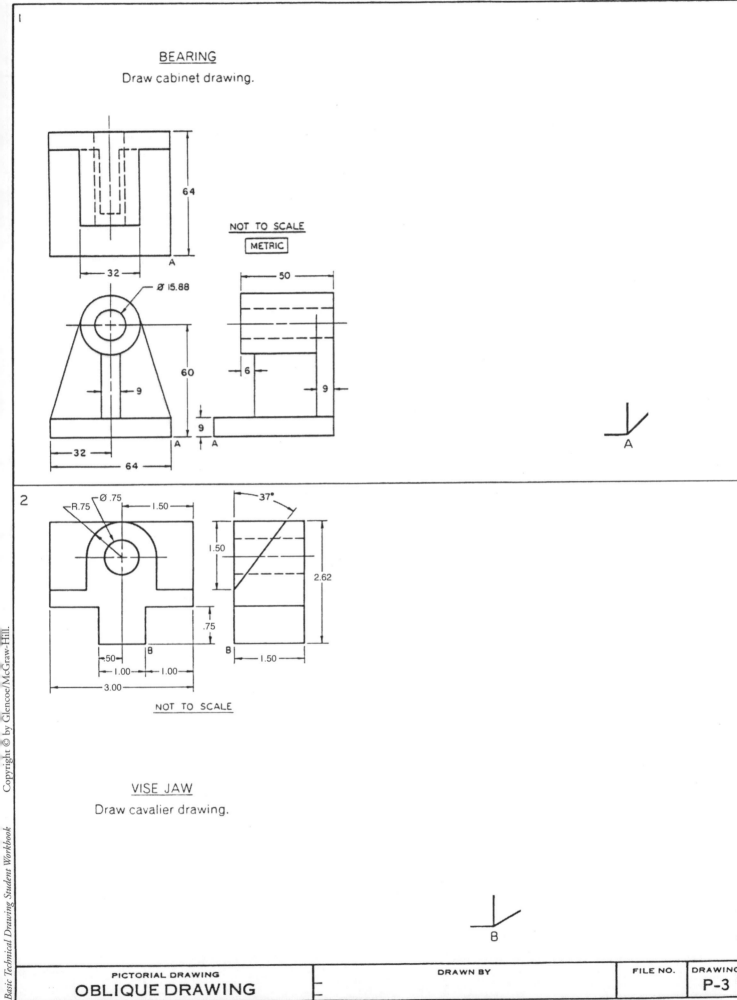

1

BEARING
Draw cabinet drawing.

64

32

A

Ø 15.88

60

9

9

NOT TO SCALE

METRIC

50

6

9

A

A

32

64

9

A

A

2

Ø .75

R.75

1.50

37°

1.50

1.50

2.62

.75

B

B

.50

1.00

1.00

1.50

3.00

NOT TO SCALE

VISE JAW
Draw cavalier drawing.

B

PICTORIAL DRAWING	DRAWN BY	FILE NO.	DRAWING
OBLIQUE DRAWING			P-3

| | DRAWN BY | FILE NO. | DRAWING |

1. Draw a development of the vertical
   surfaces of the truncated prism.

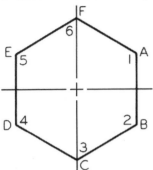

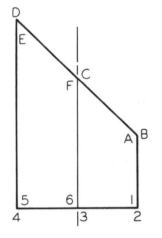

2. Draw a development of the vertical
   surface of the truncated cylinder.

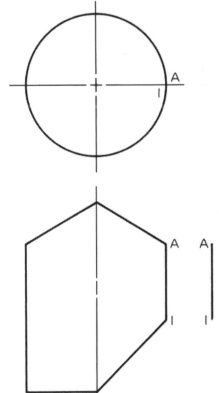

TRUNCATED PRISM AND CYLINDER	DRAWN BY	FILE NO.	DRAWING
PARALLEL LINE DEVELOPMENTS			**Q-1**

Basic Technical Drawing Student Workbook

	DRAWN BY		FILE NO.	DRAWING

1. Complete the top view and draw the
   development of the truncated pyramid.

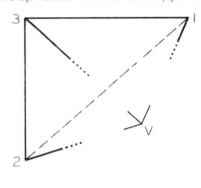

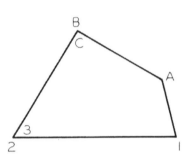

2. Draw the development of the
   truncated oblique cone.

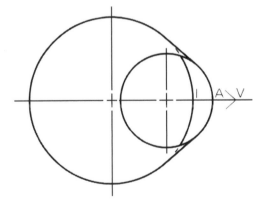

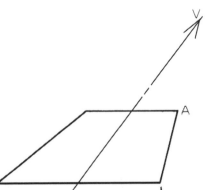

TRUNCATED PYRAMID AND OBLIQUE CONE	DRAWN BY	FILE NO.	DRAWING
RADIAL LINE DEVELOPMENTS			**Q-2**

Basic Technical Drawing Student Workbook

		DRAWN BY		FILE NO.	DRAWING

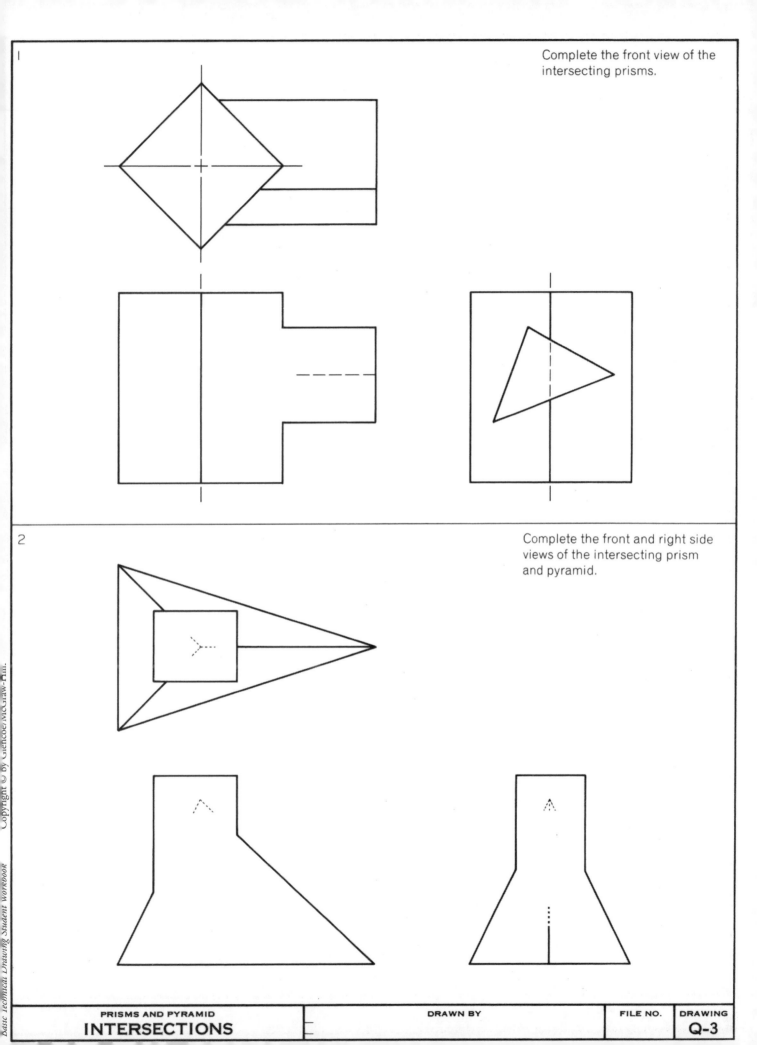

1. Complete the front view of the intersecting prisms.

2. Complete the front and right side views of the intersecting prism and pyramid.

| PRISMS AND PYRAMID | DRAWN BY | FILE NO. | DRAWING |
| INTERSECTIONS | | | Q-3 |

		DRAWN BY		FILE NO.	DRAWING

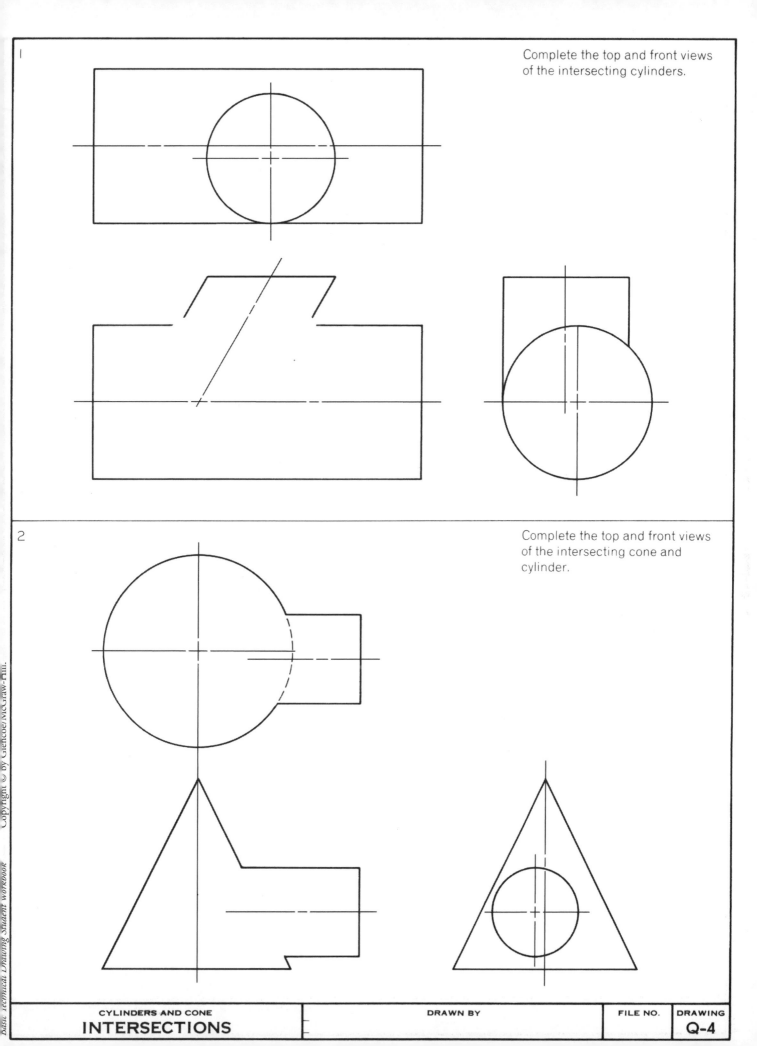

1

Complete the top and front views
of the intersecting cylinders.

2

Complete the top and front views
of the intersecting cone and
cylinder.

CYLINDERS AND CONE
INTERSECTIONS

DRAWN BY

FILE NO.

DRAWING

Q-4

| | | DRAWN BY | FILE NO. | DRAWING |

1. Draw pie chart.

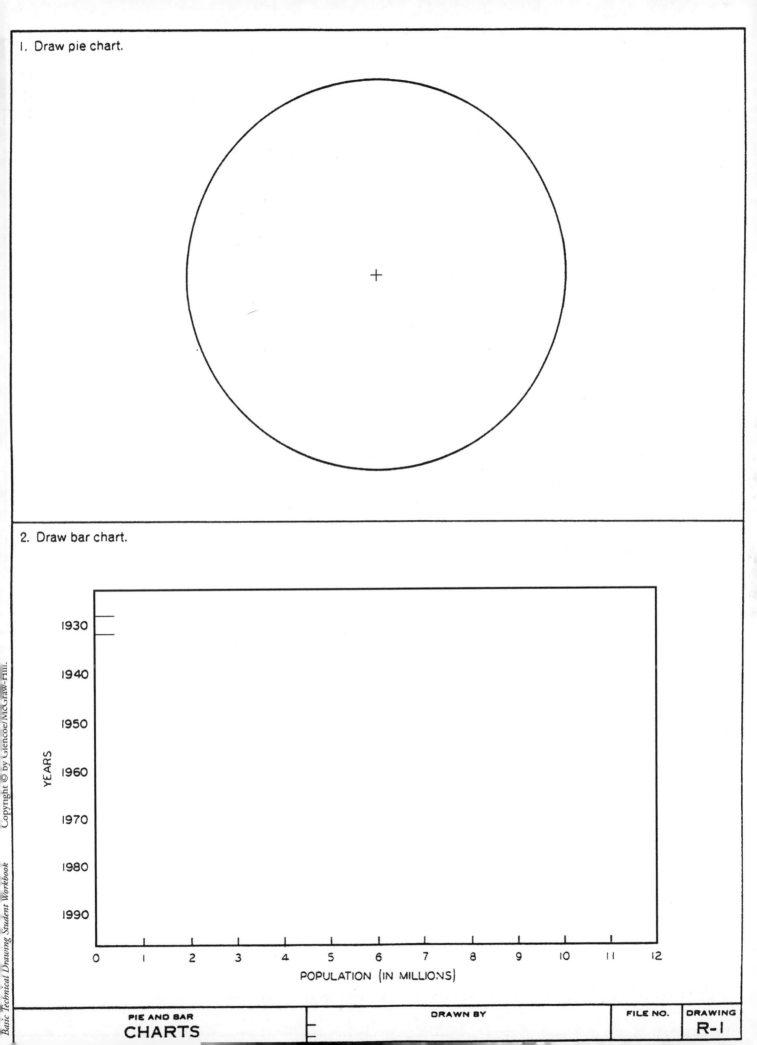

2. Draw bar chart.

YEARS

1930
1940
1950
1960
1970
1980
1990

0   1   2   3   4   5   6   7   8   9   10   11   12
POPULATION (IN MILLIONS)

| PIE AND BAR CHARTS | DRAWN BY | FILE NO. | DRAWING R-1 |

| | DRAWN BY | FILE NO. | DRAWING |

1. Draw line chart.

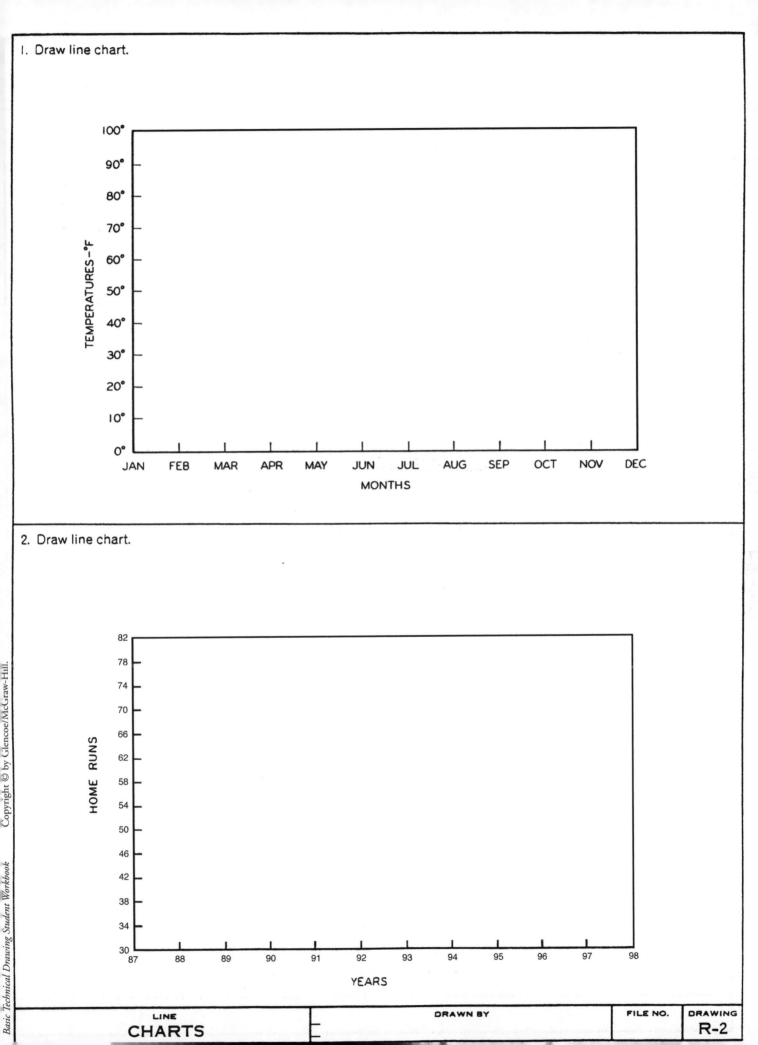

TEMPERATURES—°F

100°
90°
80°
70°
60°
50°
40°
30°
20°
10°
0°

JAN FEB MAR APR MAY JUN JUL AUG SEP OCT NOV DEC

MONTHS

2. Draw line chart.

HOME RUNS

82
78
74
70
66
62
58
54
50
46
42
38
34
30

87 88 89 90 91 92 93 94 95 96 97 98

YEARS

| LINE CHARTS | | DRAWN BY | FILE NO. | DRAWING R-2 |

Basic Technical Drawing Student Workbook

		DRAWN BY		FILE NO.	DRAWING

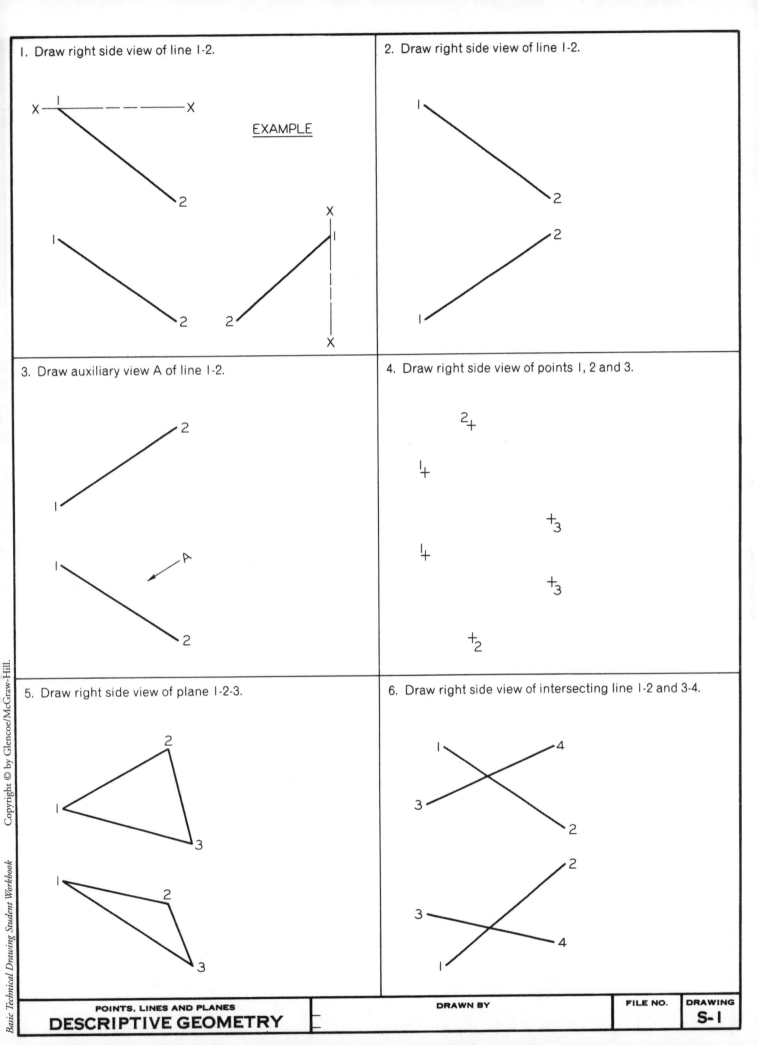

1. Draw right side view of line 1-2.

EXAMPLE

2. Draw right side view of line 1-2.

3. Draw auxiliary view A of line 1-2.

4. Draw right side view of points 1, 2 and 3.

5. Draw right side view of plane 1-2-3.

6. Draw right side view of intersecting line 1-2 and 3-4.

POINTS, LINES AND PLANES
DESCRIPTIVE GEOMETRY

DRAWN BY

FILE NO.

DRAWING
S-1

| | | DRAWN BY | | FILE NO. | DRAWING |

1. Find and measure the true length and grade of line 1-2 by drawing auxiliary view indicated.

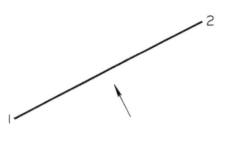

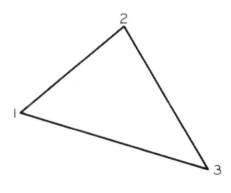

TL =   GRADE =

2. Find and measure the true length and angle with frontal plane of line 1-2 by drawing auxiliary view indicated.

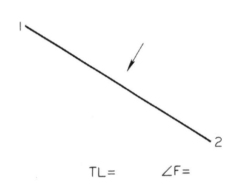

TL =   ∠F =

3. Construct a view showing the true size of plane 1-2-3.

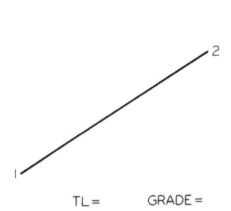

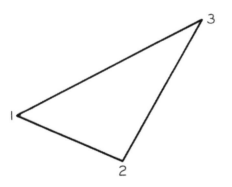

LINES AND PLANES	DRAWN BY	FILE NO.	DRAWING
DESCRIPTIVE GEOMETRY			**S-2**

DRAWN BY   **FILE NO.**   **DRAWING**

1. Find the piercing point of line 4-5 in plane 1-2-3.

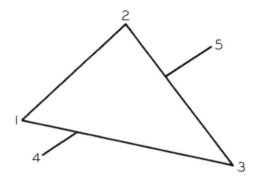

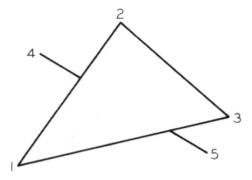

2. Determine the intersection of unlimited plane 4-5-6 and the prism.

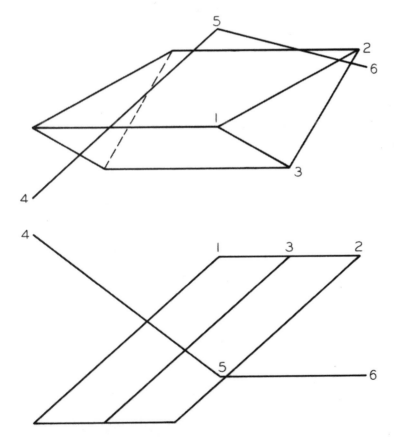

PIERCING POINTS AND PLANES	DRAWN BY	FILE NO.	DRAWING
DESCRIPTIVE GEOMETRY			**S-3**

| | | DRAWN BY | FILE NO. | DRAWING |

PART	DRAWN BY	TRACED BY	APPROVED	DATE	SCALE	FILE NO.	DRAWING

PART	DRAWN BY	TRACED BY	APPROVED	DATE	SCALE	FILE NO.	DRAWING

PART		DRAWN BY	TRACED BY	APPROVED	DATE	SCALE	FILE NO.	DRAWING

PART		DRAWN BY	TRACED BY	APPROVED	DATE	SCALE	FILE NO.	DRAWING